BENEFICENCE AND HEALTH CARE

PHILOSOPHY AND MEDICINE

Editors:

H. TRISTRAM ENGELHARDT, JR.

Kennedy Institute of Ethics, Georgetown University, Washington, D.C., U.S.A.

STUART F. SPICKER

University of Connecticut Health Center, Farmington, Connecticut, U.S.A.

VOLUME 11

BENEFICENCE AND HEALTH CARE

Edited by

EARL E. SHELP

Institute of Religion and Baylor College of Medicine, Houston, Texas

D. REIDEL PUBLISHING COMPANY

DORDRECHT: HOLLAND / BOSTON: U.S.A.

LONDON: ENGLAND

Library of Congress Cataloging in Publication Data

CIP

Main entry under title:

Beneficence and health care.

(Philosophy and medicine ; v. 11)
Bibliography: p.
Includes index.
1. Benevolence–Addresses, essays, lectures. 2. Altruism–Addresses, essays, lectures. 3. Medical care–Moral and ethical aspects–Addresses, essays, lectures. I. Shelp, Earl E., 1947– II. Series. [DNLM: 1. Ethics, Medical. 2. Religion and medicine. 3. Delivery of health care. W3 PH609 v, 11 1982 / W 50 B463]
BJ1474.B46 174′.2 82-540
ISBN 90-277-1377-4 ACR2

Published by D. Reidel Publishing Company,
P.O. Box 17, 3300 AA Dordrecht, Holland.

Sold and distributed in the U.S.A. and Canada
by Kluwer Boston Inc.,
190 Old Derby Street, Hingham, MA 02043, U.S.A.

In all other countries, sold and distributed
by Kluwer Academic Publishers Group,
P.O. Box 322, 3300 AH Dordrecht, Holland.

D. Reidel Publishing Company is a member of the Kluwer Group.

Printed in The Netherlands

TABLE OF CONTENTS

EARL E. SHELP

INTRODUCTION

The meaning and application of the principle of beneficence to issues in health care is rarely clear or certain. Although the principle is frequently employed to justify a variety of actions and inactions, very little has been done from a conceptual point of view to test its relevance to these behaviors or to explore its relationship to other moral principles that also might be called upon to guide or justify conduct. Perhaps more than any other, the principle of beneficence seems particularly appropriate to contexts of health care in which two or more parties interact from positions of relative strength and weakness, advantage and need, to pursue some perceived goal. It is among those moral principles that Tom L. Beauchamp and James F. Childress selected in their textbook on bioethics as applicable to biomedicine in general and relevant to a range of specific issues ([1], pp. 135–167). More narrowly, The National Commission for the Protection of Human Subjects of Biomedical and Behavioral Research identified beneficence as among those moral principles that have particular relevance to the conduct of research involving humans (2). Thus, the principle of beneficence is seen as pertinent to the routine delivery of health care, the discovery of new therapies, and the rationale of public policies related to health care.

Given the sort of central place the principle of beneficence commonly has in guiding and justifying activity in health care, it is important to the moral conduct of that enterprise to understand what the principle means, what it rightly warrants and what it does not. The essays in this volume respond to these concerns. A previous volume in the series, *Philosophy and Medicine* (No. 8), addressed the relevance and implications of the principle of justice for health care [3]. The essays in that collection discussed what might be owed to persons in health care as a matter of their due or right. As a companion and complement to that earlier work, these essays interpret the principle of beneficence from several philosophical and theological perspectives in order to suggest its meaning, relevance, and application to a variety of moral concerns in contemporary health care.

Much in this volume is ground-breaking. And, the contributors report that the ground was not easily broken! Several found that much of the existing theoretical and applied literature on beneficence was inadequate

Earl E. Shelp (ed.), Beneficence and Health Care, vii–xvi.

or inappropriate for the specific selected concerns addressed here. Yet, in spite of the difficulty of the assignment, all have been faithful to their charge. The creative bounty of their labors is here presented to stimulate, enrich, and advance contemporary moral reflection about the principle of beneficence and its strategic importance to health care. The essays are presented within three categories of investigation. The first section contains one historical and three philosophical studies that provide an introduction to the history of beneficence in medicine and an orientation to the foundation, nature, and function of the principle of beneficence in theories of ethics. The second section contains three essays. These examine the place of beneficence in Western religious ethics and comment on the relationship of these understandings to moral questions in health care. The third section contains four essays that investigate the meaning of beneficence for health care and the relevance of beneficence to the distribution of health care resources.

The section of background studies begins with an historical study by Darrel Amundsen and Gary Ferngren. They provide a description of a commentary on that elusive quality of the medical event that alternately has been called 'caring', 'compassion', 'humanitarianism', 'altruism', 'beneficence', and 'philanthropy'. They focus on the use of 'philanthropy' and the concept of beneficence in medical history in order to assess the meaning and implication of the concept for present health care concerns, especially the patient-physician relationship. Their review begins in the Greek period of medicine where philanthropy carried within it a notion of reciprocity or obligation on the part of the recipient to honor the giver. During the Roman period the scope of philanthropy (*humanitas*) was extended in philosophical and popular ethics to be more inclusive, giving voice to the notion of human brotherhood. But, according to the authors, it was not until Scribonius Largus (first century C. E.) that philanthropy or love of mankind was considered an essential trait of a physician. The relationship of the Christian idea of *agape* to philanthropy also is discussed. The church's interest in the cure of bodies is traced to an ethic of *agape* which was expressed at times in secular terms like *philanthropia*. As a result, hospitals were founded by the Christian church and medical charity became a prominent part of the Christian movement. This was particularly true for monasticism during the Middle Ages. With the advent of the licensing of physicians (12th century), medical philanthropy was given a different emphasis in both Christian and secular literature. The literary discussion moves to a consideration of the treatment of the poor and the setting of fees. Amundsen and Ferngren conclude that, during the periods reviewed, philanthropy was considered by some an essential feature of the true physician. For others, competence enjoyed prominence with philanthropy or compassion seen as merely desirable. The proper mix of these qualities among physicians is still subject to debate. However, the debate gains perspective with a consideration of the historical precedents provided by this essay.

The background studies continue with a discussion by Allen Buchanan of the role, function, and ground of beneficence in several moral theories. He traces the basic lineaments of each theory and gives special attention to the role of beneficence in Kant's moral theory, John Rawls's ideal contract theory of justice, utilitarianism, Robert Nozick's rights-based libertarianism, and rational egoism. His analysis draws upon three fundamental distinctions: beneficence and benevolence, beneficence in special relationships and generalized beneficence, and the respective rational foundation of special and generalized forms of beneficence. Buchanan observes that answers to questions of who are the proper recipients of beneficence in health care and the place of the principle of beneficence in a moral consideration of health care issues depend on a general philosophical theory of beneficence. Thus, he critically assesses the potential of each of the surveyed theories to provide an adequate statement and ground for beneficence, finding areas of difficulty and profit in all. He concludes his evaluation of beneficence in the selected, general moral theories with suggestions of the implications of each for moral questions in health care in terms of social policy and individual relationships.

William Frankena considers the implications of beneficence for health care from the perspective of an ethics of virtue. He contrasts an ethics of virtue with an ethics of duty. The former is concerned primarily with dispositions, motives, traits, or ways of being as morally good, independent of the actions which issue from them. The latter is concerned primarily with the rightness of actions or duty independent of certain dispositions, motives, or traits. Frankena notes that beneficence may or may not have a place within an ethics of virtue, noting that beneficence as now understood was not an important virtue in the West before Christianity, with the possible exception of the Stoics. He thinks that an adequate ethics of virtue must include justice as a cardinal virtue as well as beneficence/benevolence. He does not consider beneficence as a virtue. Rather, he combines benevolence and responsibility as the disposition that can be properly characterized as virtuous. The central component of this virtue is a motivation to do good and not evil to others. Yet, he remarks that dispositions can be good without being morally good and that virtues need not be necessarily moral ones. His representation of beneficence in an ethics of virtue as beneficence/responsibility suggests that all moral agents are to cultivate the virtue and then do in each situation what the virtue moves one to do. Frankena doubts that health care could be provided adequately in a society that relied solely on personal virtue. He discusses the implications of such an approach to various forms of individual and social relationships pertinent to health and health care. The essay concludes with a consideration of the relevance of an ethics of virtue to professional ethics.

John Reeder's review of the nature of the duty of beneficence concludes the first section. He suggests that there are three levels or forms of duties of beneficence. The first form, mutual aid, is held to be a general duty for all members of the moral community apart from institutional provisions to render assistance. Reeder distinguishes welfare and well-being, suggesting that a moral agent has a duty to promote the former at the expense of the agent's own well-being in whole or in part. He argues, recognizing that a justification of his view is problematic, that a moral agent is obligated even to sacrifice one's welfare up to an indeterminate point. This point is indeterminate because of variations in empirical predictions of the future, levels of risk people will tolerate, and value judgments regarding the ranking of various goods. Yet, he holds that such a principle of obligation would be adopted as a form of self-interest. Beyond this level of duty, Reeder's theory terms actions on behalf of others as desirable or supererogatory but not required. He suggests that supererogation can take several forms beyond the requirements of mutual aid but that, again, the difference between mutual aid and supererogation is sometimes difficult to determine. Medical treatment is not considered by Reeder as subject to mutual aid or supererogatory beneficence in situations where a specialized class of personnel function within a social institution to perform this service. In the absence of unusual circumstances, the duty of mutual aid is replaced by the system of medical care. Although his analysis of role duties for medical personnel accommodates supererogatory beneficence, "the beneficence which is proper to the therapeutic role is a duty of justice."

Transitioning from philosophical investigations, the three essays in the second section examine the role of beneficence or its equivalent in Judaism, Roman Catholicism, and selected forms of Protestantism. Ronald Green initiates this area of inquiry with a review of "Jewish Ethics and Beneficence." He finds the tradition to be rich with reflection on matters of beneficience. Opinions vary within the Jewish community regarding acts of beneficence or self-sacrificing love, depending on the weight given to the worth of each individual in a situation of conflict. Green sorts out several strands of Jewish teaching about this complex and diverse area of inquiry. He notes that the Rabbis encouraged generosity but held one's duty to be limited by one's responsibilities to oneself or one's community. Thus, the Rabbis recognized a duty to save another's life but not necessarily at the cost of one's own life. In health care contexts, the Rabbis held physicians to have certain role-related duties that required the assumption of certain risks, beyond the usual requirements, in order to serve one's patients. Green draws upon the historic views of Judaism toward self-sacrifice or beneficence to conclude his essay with a discussion of their implications for such concerns as visiting the sick, financial support of institutions of health care, donation of blood, medical experimentation, and organ donation.

William E. May finds the tradition of Roman Catholicism to be rich, like Judaism, in reflection on acts of the virtue of charity which is the equivalent to philosophical beneficence. His survey of Roman Catholic moral thought includes a report of the relevant moral teachings issued by the magisterium and an account of the debate among Roman Catholic writers about these teachings, particularly as they relate to issues in health care. The magisterium teaches that some goods like human life, justice, truth, love, and peace are good in and of themselves. Further, evil ought not be done to produce good. From these basic claims, specific norms of conduct are derived. Thus, according to May, certain acts in health care are incapable of being described as beneficent because they would violate certain principles held to be universally binding. He identifies euthanasia, abortion, contraception, and mutilation as examples of violating acts. In addition to these universal principles, May shows how the principles of 'double effect' and 'totality' have been used to interpret and apply the teachings of the magisterium to contexts of health care. The discussion then shifts to a summary of the contemporary debate within the church regarding the relationship of the principle of beneficence to the principle of utility or its Roman Catholic counterpart, proportionalism. Advocates of proportionalism hold, according to May, that there are no moral proscriptions that are absolutely unexceptional. Thus, they admit, in opposition to the view of the magisterium, that certain acts in health care can be beneficent if done for a proportionate reason. He thinks that the adequacy of this view turns on an interpretation of the principle of double effect and the distinction between a directly intended and an indirectly intended effect of some action. The major features of this debate are detailed. May concludes that the more proper view of beneficence in health care reflects agreement with the teaching of the magisterium rather than with the thought of those who favor proportionalism.

Harmon Smith examines the theologies of Martin Luther, John Calvin, and John Wesley as representative of major traditions in Protestant Christianity.

The place and meaning of beneficence in each theological system also is indicated. Smith reports that Luther held that doing good works is a human possibility that comes from God. Beneficence, or God's gift of love to humanity, is reserved for God. Based upon his anthropology, Luther thought that the possibilities for human beneficence were limited. Calvin's view of human beneficence, according to Smith, also grows out of his doctrines of mankind and God. Calvin held that the atonement of Christ which makes possible human beneficence is efficacious only for those preordained by God to eternal life. Smith observes that Calvin's theology warrants 'works' without providing grounds for 'good works' or beneficence. Following Calvin and Luther by approximately 200 years, John Wesley's emphasis was not on salvation alone but on the implications of salvation for the daily lives of believers, i.e., that there is more to the Christian life than justification or salvation. Thus, Wesley departed from Luther and Calvin to allow for the possibility of relatively good acts because humanity is thought never to be totally separated from God due to God's prevenient grace. However, Wesley thought that truly good works are possible only after justification, since only then are they done as God wills them. Smith thinks that Luther's and Calvin's theologies do not provide a viable concept of beneficence but that Wesley's does. Smith's analysis moves to the American expression of Protestantism to illustrate how and why American Protestantism became so concerned with promoting the well-being of the neighbor. This trait is attributed to the influence of the Reformation in Europe and the Enlightenment in England. The effect was to detach morality from the religion of the American Protestant experience. The operational characteristics of American Protestantism became individualism, autonomy, and pluralism. Smith suggests that American Protestantism basically has not affected American health care. He notes the active participation of Protestant ethicists in the contemporary discussion of medical ethics, but observes that it is rare for one to ground one's views in theological or ecclesiastical authority. Rather, the opinions and suggestions within this literature parrallel those of conventional wisdom. Smith laments that issues in health care seem irresolvable because there is no agreement on the goal or goals of human moral agency. One senses that Smith thinks Protestant ethics in its American form does not provide a significant resource for settling these issues.

The last section of essays addresses more specifically the application of beneficence to health care in general. Natalie Abrams's essay begins the examination of benefience in health care by exploring the scope and boundary of the duties of beneficence in this setting. She focuses on the patient-health professional relationship emphasizing the duties of the professional. Beneficent conduct by health care providers is characterized as behavior that is intended to be helpful, is done independent of reward to the agent, and benefits the recipient. Among the possible sources of duties of beneficence in health care, Abrams prefers implied or expressed agreements over notions of reciprocity or need. She reviews definitions of the scope of the duty of beneficence offered by Peter Singer, Michael Slote, Tom Beauchamp, and James Childress and finds that they all consider the relative risk and harm to the agent and beneficiary as limiting factors. What each fails to mention is a regard for the wishes of the recipient as a limiting factor. In addition, ignorance, availability of resources, and competing rights and interests are seen as limitations. Abrams concludes that the traditional view that limits the scope of beneficence to an individual patient is both practical and respectful of individual rights.

The next two essays consider the meaning and application of the principle of beneficence to health care. Earl Shelp analyzes some of the variables and qualities of the principle that are necessary to its expression. His discussion is limited to the patient-physician relationship. He suggests that the potential for diverse understandings of beneficence is indicated by the range of formulations of its proper end or goal. Each of these is seen as basically formal and perhaps a form of the good, benefit, or end that the principle intends. It seems to Shelp that the principle provides a general policy for conduct without specifying the content or means of the policy. The prospect for diverse interpretations is further illustrated by a discussion of the difficulty in arriving at a uniform understanding of the principle's definitional elements of 'good' and 'harm'. Further, he explores how these diverse understandings are subject to factual, perceptual, and evaluative influences. The analysis of the principle concludes with a discussion of three selected qualities or conditions of beneficence that enhance the moral value of benefitting actions. How these qualities of respect for persons, regard for self-esteem, and value of freedom and liberty are understood, weighted, and expressed are held to influence the character of beneficence. Shelp suggests that these are necessary conditions of beneficence which help to distinguish it from alternate forms of other-regarding activity. He concludes that the principle does not admit to only one interpretation. Similarly, its application to complex situations can take many forms. This is especially true in health care where goods and harms are also subject to diverse understandings. Thus, the potential for disagreement regarding the relevance and application of beneficence to health care seems limitless. Given these conditions, he suggests that forms of beneficence in health care must be adapted to unique circumstances.

A study of the implications of the principle for health policy is provided in the essay by James Childress. His discussion is focused on the relevance of beneficence to those governmental policies intended to prevent ill health and early death. More particularly, he is concerned with that form of prevention that seeks to change individual life-styles and behavioral patterns that contribute to morbidity and mortality. Interventions into personal life-styles in order to reduce risk-taking have been based on the principle of social harm, the principle of paternalism, or some conbination of the two with both invoking beneficence. Childress examines the relevance of these principles to the regulation of life-styles that affect health. He concludes that governmental intervention has its greatest warrant by beneficence when it is aimed at the reduction of nonvoluntary or involuntary risks of ill health or early death. He is less certain about interventions with çonduct that is voluntary and self-regarding because of the lack of supporting empirical data.

Another facet of health policy is explored by Ronald Green's commentary on the place and function of altruism in a system of health care which is grounded in notions of rights or claims. He describes the traditional links between altruism and health care. On the one hand the link is philanthropic impulse which prompts care for the sick. On the other hand the link is the development of human compassion which is nurtured by caring for the sick. These links seem threatened by the changing context and dynamics of modern health care in which person-to-person relationships are interfered with by technology. Others suggest that the link is weakened further by the assertion that health care should be subject to the many individual claims upon social resources. And still others, challenging the link even more, question the purity of the motives of providers of health care. Green describes how efforts to preserve altruism in health care and yet free it from possible personal abuse have resulted in various forms of institutional or social intervention in the traditional altruist-recipient relationship. Thus the new form of medical altruism is the expression of concern through impersonal institutions like Medicaid and Medicare. He does not find the objections to this impersonal form of altruism in health care sufficient to displace it. Rather, the objections constitute new kinds of moral problems or challenges that must be addressed within a restructured or differently grounded health care system. Green concludes that, even in a health care system based on rights, opportunities for personal altruism should always be encouraged and would be desirable.

It is clear that this collection will not be the final word on the meaning and application of the principle of beneficence to moral questions in health care. The fact that it is nearly the beginning word indicates that much serious investigation and reflection needs to be done. For those who continue this investigation, these essays should make the work a little less difficult but by no means will it be made easy. The relative inattention given to the moral principle of beneficence, commonly considered central to the moral life in general and to health care in particular, deserves to be overcome. No one who has contributed to this publication would claim anything other than that a word, perhaps only a tentative word, on a vital topic has been spoken. It is surprising to find this omission in the philosophical literature. It is also surprising that the traditions of Protestant Christianity are unable to contribute much to the task. Perhaps the vitality of the general discussions within Judaism and Roman Catholicism can provide guidance and insight for the investigations that are urgently indicated. A broadened scholarly investigation, both philosophical and theological, may result in an enriched interpretation of the principle of beneficence that is able to transcend the respective boundaries of scholarly traditions and thereby provide a bridge for understanding and instruction in the moral life.

Bringing a volume to completion is always a happy event. Part of the joy comes from being able to thank those many individuals who have had an important part in its development and production. As editor, I am indebted to each of them and grateful to each for his or her skillful and thorough completion of his or her assignment. The assistance of the editors of the series, H. Tristram Engelhardt, Jr., and Stuart F. Spicker, has been valuable. Their counsel, along with the skillful assistance of Susan M. Engelhardt, during the preparation of the volume, is greatly appreciated. Mrs. Audrey Laymance deserves special mention. She worked tirelessly as editorial assistant in the preparation and typing of the manuscript. Finally, my debt to Ronald H. Sunderland is gratefully acknowledged. He is not only a colleague whose energies are an inspiration, but a friend whose encouragement during times of fatigue and disappointment provides the incentive to continue worthy tasks. Without the commitment of these individuals, this volume would still be merely an unfulfilled hope. I thank each of them for their role in helping it become a reality.

September, 1981

BIBLIOGRAPHY

1. Beauchamp, T. L. and Childress, J. F.: 1979, *Principles of Biomedical Ethics*, Oxford University Press, New York.
2. National Commission for the Protection of Human Subjects of Biomedical and Behavioral Research: 1978, *The Belmont Report: Ethical Principles and Guidelines for the Protection of Human Subjects of Research*, Washington, D. C., DHEW Publication No. (OS) 78–0012.
3. Shelp, E. E. (ed.): 1981, *Justice and Health Care*, D. Reidel Publ. Co., Dordrecht, Holland; Boston, Mass. U.S.A.

SECTION I

HISTORICAL AND CONCEPTUAL BACKGROUND

DARREL W. AMUNDSEN AND GARY B. FERNGREN

PHILANTHROPY IN MEDICINE: SOME HISTORICAL PERSPECTIVES

There is a somewhat nebulous and elusive quality usually desired and sometimes present in medical practice, variously called caring, compassion, humanitarianism, altruism, beneficence, or philanthropy. None of these words individually is adequate to encompass the quality to which this historical investigation is addressed, and none is truly synonymous with any other. Each of these words can be taken as having a negative connotation in that each assumes, as it were, even if only slightly, a need or a deficiency in someone to which the quality expressed by the word is, at least in part, a response. As one reads through this list of words, that negative connotation grows in intensity until, when one reaches 'philanthropy', one is dealing with a term that has of late, when applied to medical practice, received some rather 'bad press'. This is especially evident in articles by William F. May [31] and Robert M. Veatch [45], both of whom are concerned with some fundamental questions of the basis for medical ethics and obligations within the physician/patient relationship. Both stress the desirability of obligations founded on reciprocity rather than on deontology that arises from the profession's own definition of its role-specific duties. Both authors consider philanthropy as a central feature of the latter, that is, of a one-sided definition of obligations. May writes about "the conceit of philanthropy when it is assumed that the professional's commitment to his fellow man is a gratuitous, rather than a responsive or reciprocal, act. Statements of medical ethics that obscure the doctor's prior indebtedness to the community are tainted with the odor of condescension" ([31], p. 31). There is a degree of condescension, or at least a potential for suspicion of condescension, in each of the words listed above, indeed in that very quality in medical practice for which we are at a loss to find a suitable word. In the present paper we shall avoid being prescriptive as we seek to describe that elusive element in the history of ancient and medieval medicine.

"For where there is love of man [*philanthropia*]," reads a famous passage in the pseudo-Hippocratic treatise *Precepts* (*Precepts*, 6), "there is also love of the art [*philotechnia*]."[1] This lofty sentiment, removed from its context, has often been taken to affirm the existence of a genuine philanthropic impulse in Greek medicine. Thus Sir William Osler saw in this maxim evidence

Earl E. Shelp (ed.), Beneficence and Health Care, 1–31.

of the Greek physician's "love of humanity associated with the love of his craft – *philanthropia* and *philotechnia* – the joy of working joined in each one to a true love of his brother."[2] More recently, P. Lain-Entralgo construed the adage as evidence that the Greek physician's relation with his patient "consisted in a correct combination of *philanthropia* (love of man as such) and *philotechnia* (love of the art of healing). A doctor was thus a friend to his patient both as 'technophile,' or friend of medicine, and 'anthrophile,' or friend of man." He further maintained that "a careful study of the Hippocratic writings leads to the conclusion that Hippocrates and his direct and indirect followers were 'philanthropists' *avant la lettre*" ([15], pp. 21f and 245, n. 1). In order to gain a reasonably accurate sense of the place of philanthropy in ancient medicine, let us begin by briefly considering the meaning of *philanthropia* and the concept of beneficence in classical antiquity.

The term 'philanthropy' is derived from the Greek word *philanthropia,* which means literally 'love of mankind'.[3] The original meaning of the word was the benevolence of the gods for man, a concern that manifests itself in the granting of gifts and benefits. By a natural extension of meaning the word came to refer as well to the munificence and generosity of rulers toward their subjects and to the friendly relations between citizens and states. In all these meanings we find common elements of condescension and the giving of gifts or benefits that the word never lost. In the fourth century before Christ, the word came to be used with the more general meaning of 'kindly, friendly, genial' in reference to personal and social relationships. It is widely used in this sense, e.g., in the Hippocratic Corpus, to indicate a kindliness, courtesy, and decent feeling towards others. It is doubtless with this meaning in mind that Diogenes Laertius says that *philanthropia* may take three forms: that of salutation, of assisting one in distress, and of fondness in giving dinners. "Thus philanthropy is shown either by a courteous address, or by conferring benefits, or by hospitality and the promotion of social intercourse" (*Lives of Eminent Philosophers*, 3, 98). In the Hellenistic period the word takes on a much more comprehensive meaning and is sometimes used to express a love of humanity, of one's fellow man generally, suggesting a fëeling of concern for the well-being of one's fellows. Yet even in this sense *philanthropia* continued to retain its original meaning of a relationship between a social superior and inferior, a condescending benevolence, which reflected the limitation in the classical world of the philanthropic impulse.

In general it may be said that philanthropy among the Greeks did not take the form of private charity or of a personal concern for those in need, such as orphans, widows, or the sick.[4] There was no religious or ethical impulse for almsgiving; nor was pity recognized either as a desirable emotional response to need and suffering or as a motive for charity. By contrast with the emphasis in Judaism on God as particularly concerned for the welfare of the poor, the Greek and Roman gods showed little pity on them; rather they showed greater regard for the powerful, who could offer them sacrifices. Pity as an emotion was reserved, not for the indigent, but for those (mostly members of the upper classes) who had experienced a reversal of fortune that had reduced them to poverty; because the lower classes had never experienced a catastrophic fall they could not deserve pity. As a motive for assisting those in need, pity was shown by those who, on the one hand, could sympathize with members of their own class in need and, on the other, might hope to build up a fund of good will in case they should experience a similar misfortune. The Stoics regarded pity "not as a liberating emotion necessary to inspire the selfless service of others, but as an emotion which enslaved a man's mind and spirit, and undermined the good man's claim to self-sufficiency and self-command" ([21], p. 81). The basis of generosity or of any moral action for a Stoic should be rational rather than emotional; the latter was regarded as impulsive and subjective, the former as objective, universal, and humanitarian. This is typically the classical view.

It was only on a *quid-pro-quo* basis that pity might serve as a motive for giving: the giver hoped that, should he ever be in need, he might expect pity and aid because he had earned it by displaying pity himself. Hence pity might most properly be felt for the members of one's own class, from whom reciprocation could be expected. When it was shown more generally, it was out of an instinctive sympathy for the human condition, as in the *arai Bouzygeiai*, "curses which were called down upon any man who failed to provide water for the thirsty, fire for anyone in need of it, burial for an unburied corpse, or directions for a lost traveller" ([21], p. 46). The motivation for such acts is to be found in the statement attributed to Aristotle, "I gave not to the man, but to mankind" (Diogenes Laertius, *Lives of Eminent Philosophers*, 5, 21). Here benevolence is a form of hospitality rather than of justice or moral or religious obligation. One is to do as one would be done by. Even in such simple acts of human concern, there was an eye to reciprocity. One might someday require similar assistance, perhaps even from the person one had helped in the past.

It is in this light that one should approach the dictum in the *Precepts* with which we began. The statement, "where there is love of man [*philanthropia*], there is also love of the art [*philotechnia*]," appears in the middle of a paragraph dealing with the question of medical fees. The opening sentence begins with the admonition, "I urge you not to be too unkind." The word translated "unkind" is the noun *apanthropia*, which is essentially an antonym for *philanthropia*.[5] The physician is thus urged not to be too 'unphilanthropic' but to consider his patients' financial means and to treat gratuitously the stranger in financial straits. "For where there is love of man there is also love of the art" is often interpreted to mean that when the physician is a lover of man he will be, in response to that sentiment, also a lover of his art. We should then expect it to be followed by the conclusion that a physician's being thus motivated by both his *philanthropia* and his *philotechnia* will lead to his extending compassionate care to his patients. But instead we find that the *philanthropia* belongs to the physician and the *philotechnia* to the patient: "For some patients, though conscious that their condition is perilous, recover their health simply through their contentment with the goodness [*epieikeia*, clemency, natural mildness] of the physician" ([23], vol. 1, p. 319). Thus the patient's response to the physician's *philanthropia* takes the form of *philotechnia*, love of the physician's art, which reveals a contentment with the physician's *epieikeia* or kindness. This contentment greatly aids in the curative process. The *philanthropia* of the physician here seems to be little more than kindliness and charitableness.

A very similar interpretation must be given to a passage in the pseudo-Hippocratic work *On the Physician* (*On the Physician*, 1).[6] Here the concern is with the proper deportment that a physician's dignity requires. In the context of much sage advice on medical etiquette and morality appears the statement that the physician "must be a gentleman" . . . who is "grave and *philanthropos*." The last word is the adjective derived from *philanthropia*. Is the Greek physician being urged here to be a "lover of mankind" as a motive for his practice? To this question an unequivocal 'no' must be given. W. H. S. Jones nicely captures the meaning of *philanthropos* here when he translates it in this passage as "kind to all" ([23], vol. 2, p. 311). A few sentences later the physician is urged not to appear harsh, for then he would seem to be *misanthropos*, which here probably means little more than "unkind".

In both passages *philanthropia* seems to designate "a proper behavior toward those with whom the physician comes in contact during treatment; it is viewed as a minor social virtue . . . " ([14], pp. 321f). Hence it can be little more than a guide, not an impulse or motivation, for the practice of medicine. Rather the motivation of the classical physician to practice medicine seems more often to have been *philotimia* rather than *philanthropia*.

The idea of reciprocity in philanthropic giving was basic to the classical world. Every benefit conferred, every gift, philanthropic or charitable, obligated the recipient to some form of repayment. This exchange of giving, which is found as early as Homer, may well have had its origin in the context of an aristocratic society of equals for whom giving and counter-giving cemented friendships. Such a relationship, which brought advantages to both parties, eventually spread beyond the upper classes and came to involve, to some degree, all the members of a community. It is, e.g., the basis of the patron-client relationship in Rome, and it came to involve services (*beneficia*) rather than merely material gifts. It came as well to include the relationship between unequals, such as we find in the word *philanthropia*. Where this association existed between the wealthy and the poor, the only return that could be made by the poor was 'honor' in the form of public or private recognition of the philanthropy of the benefactor. The desire for honor and public recognition served as one of the chief motives of personal behavior in the classical world. Public philanthropy was one of the most important means of obtaining honor in the community and it was by no means uncommon for benefactors to admit that they were giving in return for public recognition. "The Greeks, in particular, believed that the good man would pursue honor, admiring as they did a strong competitive element in man's psychology . . . " ([21], p. 43). "It is quite clear," writes Cicero, "that most people are generous in their gifts not so much by natural inclination as by reason of the lure of honor – they simply want to be seen as beneficent" (*De Officiis*, 1, 14, 44; [21], p. 47).

The return of honor for a benefaction had special reference to the operation of the city-state in the Graeco-Roman world, where many of the financial burdens of the community were met by the wealthy class either by the holding of a public office that often required considerable personal expense (liturgy) or by an appeal for a public subscription to the wealthy. It was a regular practice to obtain much of the public revenue of a city from the gifts of the wealthy; in the case of a subscription a motion would be made to establish a fund for a need, to which the wealthy members of the community were expected to contribute. The impulse for such giving was, positively, *philotimia* ('love of honor') or *philodoxia* ('love of glory'); and, negatively, the threat that the wealthy might be exposed to prosecutions that might result in the loss of either their position of honor or their wealth. As a return for a subscription the community often rewarded wealthy benefactors by setting up honorary inscriptions, which recorded, often in great detail, on stone or bronze, the nature and amount of the benefaction. Thousands of these inscriptions remain today that testify to the public philanthropy of the wealthy – and others, such as physicians, teachers and philosophers

– who made public benefactions or performed some public service. Such benefactions were not made out of a disinterested 'love of mankind' but rather out of a desire for personal recognition, which the Greeks and Romans regarded as the natural motive for giving.

The impulse for giving was not pity: "Broadly speaking, pity for the poor had little place in the normal Greek character, and consequently for the poor, as such, no provision usually existed; the idea of democracy and equality was so strong that anything done must be done for all alike; there was nothing corresponding to our mass or privately organised charities and hospitals" ([41], p. 110). Hence when gifts were made or services performed, they were for the entire community, no distinction being made between the destitute and others. The basis for philanthropic giving was the possibility of reciprocity rooted in the idea of the friendship and cooperation that bound together the members of the city-state. The destitute could not return the favor of a benefaction; hence they were not deemed worthy as a class of receiving special aid. Such philanthropy was civic, not personal, intended for the community, and therefore limited to the citizens of the community on an equal basis. Outsiders were excluded, for they had no moral claim on a public benefaction. Rather than pity it was the cooperation between all citizens of the state, rich and poor alike, that provided the moral claim of all citizens to the public benefaction. Had pity been the motive, the citizens would have lacked any legitimate claim to the gift, for they would not have been able to reciprocate by honoring the benefactor. Hence, because of the *quid-pro-quo* basis of the philanthropy, pity was rigidly excluded as an improper motive; giving motivated by pity would have been resented, for the receiver would be unable to reciprocate ([21], p. 85).

It is in this context that the numerous public inscriptions honoring physicians must be understood. "Nothing leaves a more pleasing impression," write Tarn and Griffith, "than the numerous decrees of thanks passed to physicians" ([41], p. 109).[7] The physicians so honored are described as tireless in their services on behalf of the community, devoted to their profession, making themselves available to all who need their services, serving rich and poor, citizen and foreigner alike, remitting fees, remaining in the city during an epidemic. Here, if anywhere, there seems to be *prima facie* evidence for a genuinely disinterested 'love of mankind' as a motive for service. Yet there is nothing to distinguish honorary decrees for physicians from the whole class of honorary decrees passed by Greek cities for benefactors of all kinds. The language is formulaic and the benefactions for which physicians are honored can be at least partially paralleled elsewhere. Thus if physicians sometimes remitted fees for those unable to pay, so did philosophers on occasion, and they too were honored by public decrees.[8] Physicians were rewarded for their service to the community in the ordinary way in which communities rewarded benefactors: by public honors voted them. Hands quotes the pseudo-Hippocratic *Maxims* as indicating that for the physician money is of secondary importance to honor: "the quickness of the disease . . . spurs on the good doctor not to seek his profit, but rather to lay hold on reputation" (*Maxims*, 4, 6; [21], p. 131). The honorary inscriptions suggest that *philotimia* was an important, if not the chief, motivation of many classical physicians.

Henry Sigerist wrote that "every period has an ideal physician in mind, indeed must have one . . . "([37], p. 273). The ideal physician and the physician as an ideal are types encountered with frequency in classical literature. The word 'physician', used figuratively, was not a neutral term. Unless it was modified by a pejorative adjective, it usually carried the metaphorical force of 'a compassionate, objective, unselfish man, dedicated to his responsibilities.' Thus, for example, the good ruler, legislator, or statesman was sometimes called the physician of the state. Essentially, the statesman is (or should be) to the state what the physician is to his patient.[9] Similarly, ancient philosophers sometimes considered themselves physicians of the soul. Regardless of whether the 'medicine' they administered was soothing or painful, it was their 'patients'' good that was always their proper object.[10]

Irrespective of how far short of the ideal many physicians may have fallen, an ideal did exist. Was a physician considered to be a physician only insofar as he lived up to such an ideal? Naturally most people, laymen and physicians alike, would never have asked such a question; but some writers did. Plato, for example, in the *Republic* (*Republic*, 340 C–347 A) discusses the question whether self-interest is the motive behind all human efforts, especially political activity. A comparison is made between politics and various arts, including medicine. In this context the question is asked whether the physician *qua* physician is a healer or an earner. *Qua* physician he is exclusively a healer since in that capacity his interest is entirely in providing the advantage for which his art exists. Acting *qua* physician he does not seek his own or his art's advantage, but only his patient's. His earning of an income or his gaining of honor from his art is itself a subsidiary art and follows from the practice of his primary art. Thus the motivation for practicing any art, whether it be for money or honor, is quite irrelevant to the integrity of the art itself, since the *raison d'être* of the art is the furnishing of the good for which it was created. Motivation is not the issue; competence in the art is what is essential, for without competence the putative practitioner of any art fails in fact to be a practitioner owing to his incompetence to achieve those ends for which his art exists.

Galen, in a work entitled *On the Doctrines of Hippocrates and Plato*[11], discusses this specific passage from the *Republic*. After summarizing the argument, he writes, "Some pursue the medical art for the sake of money, others for the exemptions granted by law,[12] certain ones on account of *philanthropia*, just as others for the glory or honor attached to the art." Galen goes on to say that they are all called physicians, insofar as they provide health; but insofar as they are led by different motives, "this one is a *philanthropos*, that one a lover of honor, another a lover of glory, and yet another a seeker after money." Therefore, the aim of physicians *qua* physicians is neither glory nor reward, as the Empiricist Menodotus[13] wrote. Such, however, was Menodotus' motive, although it was not the motive for Diocles,[14] just as it also was not for Hippocrates and Empedocles[15] or for many of the physicians of old who treated men because of *philanthropia*." Galen's understanding of the force of the words *philanthropia* and *philanthropos* seems to be different from that of the authors of the pseudo-Hippocratic treatises, *Precepts* and *On the Physician*.

The reason for this may lie in the influence of humanitarian and cosmopolitan ideas on both philosophical and popular ethics in Hellenistic and Roman thought. As has already been observed, after the fourth century before Christ the word *philanthropia* came to be used to express a comprehensive love of mankind, a common feeling of humanity. It has been suggested that this change is due to Alexander's conquest of the East; or perhaps it was an inevitable result of the lessening importance of the polis and the growing individualism of the fourth century.[16] In any event, this theme of a common kinship of mankind was taken up by Cynicism and Stoicism and influenced both philosophical and popular ethics in Hellenistic Greece and the early Roman Empire.[17] One finds in the writings of the Stoics particularly, in Musonius Rufus, Seneca, Epictetus, and Marcus Aurelius, an emphasis on the brotherhood of all men, a love of one's enemy, and forgiveness of those who have done wrong to us. Philosophy was regarded by the educated as the moral educator of humanity, and it is apparent that cosmopolitan and humane ideas of Stoicism influenced Roman society, in, for example, an increasing humaneness in the attitude toward slavery. Stoic emphasis on cosmopolitanism and human brotherhood seems to have influenced the concept of *philanthropia*, which is used to denote humane and civilized feeling towards humanity in the Roman Empire in the sense in which Galen seems to use the term.[18] Aulus Gellius reflects this meaning when he says that the Latin word *humanitas* is commonly taken to have the same meaning as the Greek word *philanthropia*, which signifies "a kind of friendly spirit and good-feeling towards all men without distinction" (*Attic Nights*, 13, 17, 1). Edelstein believes that "philanthropy became integrated into the ethical teaching of the dogmatic physicians not long before Galen's time, if indeed it was not Galen himself who accepted the ideal of philanthropy in accordance with his Stoic leanings" ([14], p. 336, n. 29). There can be little doubt that Galen reflects the greater humanitarianism that was taught by the Stoicism of his day.

While there was no exact equivalent of *philanthropia* in Latin, there was a word that came to have many of the same associations: *humanitas*.[19] Aulus Gellius, in the passage cited above, writes that the word has the force of the Greek *paideia*, "what we call education and training in the liberal arts." *Humanitas* comprehends the humane virtues that we expect of an educated person: politeness, tolerance, command of the social graces; but also kindliness, mercy, consideration of others. The word is a favorite of Cicero, who defined those qualities that he believed a liberal education should produce in a person. It is not surprising that by the time of Gellius *humanitas* should have come to be synonymous with *philanthropia*. It has been suggested that the Roman concept of *humanitas* goes even beyond the Greek concept of *philanthropia*, that "it may have conveyed the idea of a warm, human sympathy for the weak and helpless in a measure which *philanthropia* never did" ([21], p. 87). This is unlikely, for the concept of *humanitas* was limited to a narrow circle of urban and educated aristocrats. Nevertheless, the word reflected the qualities that were expected to characterize the Roman ruling class and which motivated much of the humane legislation of the early Empire.

Edelstein thus summarizes the fundamental change in medical deontology which he believes resulted from these changes which were occurring in the underlying philosophical systems at this time: "The morality of outward performance characteristic of the classical era was now supplemented by a morality of inner intention. The physician – whether as an amateur philosopher or as a philosopher in his own right – had learned to regard his patient not only as the object of his art, but also as a fellow man to whom he owed more than knowledge alone, however great, can provide" ([14], p. 335). It is very unlikely that the average physician of that period could be considered even "an amateur philosopher", much less "a philosopher in his own right." Galen, however, can quite properly be placed at least in the former, and perhaps even in the latter, category. Possibly of all his works, there is none that is more fundamental for one's understanding of him than the short treatise entitled *That the Best Physician is Also a Philosopher* (*Quod optimus medicus sit quoque philosophus*, [19], vol. 1, pp. 53–63). It is Galen's foundational principle that both medical research and treatment must be based on philosophy and therefore the best physician must also be a philosopher. Galen did not limit the role of philosophy in medicine simply to supplying the scientific framework that natural philosophy provides, but insisted that philosophy provide the ethical principles for medical theory and practice as well. Thus the best physician must be a philosopher and as a philosopher he must be "self-controlled and just and immune to the temptations of pleasure and money; he must embody all the different characteristics of the moral life which are by their very nature interdependent" ([46], p. 82). A predominant feature of this 'moral life' for Galen was *philanthropia*. This *philanthropia* manifested itself in his claim that he never demanded remuneration from any of his pupils or patients. He further tells us that he often provided for the various needs of his poorer patients.[20] In doing this he felt that he was following the example of the ancients, particularly of Hippocrates. Hippocrates – an idealized and nearly mythical Hippocrates – was for Galen an exemplar of medical probity and virtue. In this treatise Galen says that Hippocrates turned down the lucrative position of physician to a powerful Persian satrap in order to stay in Greece and take care of the poor. A little earlier in the same work, Galen calls medicine "an especially philanthropic art."[21] Hippocrates' spurning money[22] and choosing to treat the poor freely is not the only evidence of his philanthropy cited to show that medicine is "an especially philanthropic art." He adduces as additional evidence the facts that Hippocrates travelled about "to verify by experience what reasoning had already taught him about the nature of localities and waters" ([42], p. 63), and that Hippocrates published his medical knowledge for the good of mankind.

For Galen medicine was "an especially philanthropic art" for two major reasons. (1) Regardless of whether or not *philanthropia* provided the motive for any particular physician, the art itself when practiced by a competent physician was a philanthropic art in that it relieved mankind's sufferings. (2) If the physician was motivated by *philanthropia*, which he would be if he was also a philosopher, he would demonstrate his philanthropy in the ways in which both Hippocrates and Galen did: compassionate care of the destitute; advancement of medical knowledge; and dissemination of that knowledge to contemporaries and for posterity. In Temkin's words, philanthropy is for

Galen "the love of mankind . . . and the concern for its future . . . . Galen's philanthropy is not only that of the physician, but more comprehensively that of a philosopher who subjectively delights in study and objectively labors for the good of mankind" ([42], pp. 49f).

Narrower in scope but deeper in its emphasis on compassion than Galen's view of philanthropy were the ideas espoused by Scribonius Largus, a Roman physician who lived in the first century of the Christian era. Scribonius' one surviving work is his treatise *On Remedies*, which he prefaced with a short essay bearing the title *Professio medici*,[23] which, as the title indicates, deals with the profession or calling of the physician. Neither of these is an adequate translation of *professio*, which in Scribonius' time was a word charged with Stoic overtones. Panaetius, a Stoic philosopher of the second century B. C., whose influence on Cicero permeates the latter's *De officiis*, maintained that any legitimate role that one assumed had central to it various duties (*officia*). If one who occupies a particular role is faithful to the *officia* inherent in that role, one is acting morally and justly as regards one's *professio*. If one is unfaithful to the *officia* of one's role, one is not only acting immorally but is violating the integrity of one's *professio*. When violating the *officia* of one's *professio* one ceases to occupy that *professio* and is no longer, or at least is not then, (for example) a judge, a lawyer, or a physician.

Scribonius was a proponent of drug-therapy, an adversary of physicians who rejected drugs and employed only dietetics. He writes that some of these do so out of ignorance, which is reprehensible. Others, who know the usefulness of drug-therapy, deny it to their patients out of jealousy toward their colleagues who treat their patients more effectively. Such physicians are even more to be condemned than those who are simply ignorant. "They ought to be despised by gods and men, all those physicians whose heart is not full of compassion [*misericordia*] and humaneness [*humanitas*] consonant with the will [*voluntas*] of the *professio* itself." Therefore the physician will harm no one, and "because medicine does not have regard for men's circumstances or their character, she [*medicina*] will promise her succor equally to all who seek her help and she promises never to harm anyone." Citing the so-called Hippocratic oath, he asserts that Hippocrates, in forbidding the practice of abortion, "had come a long way in the direction of preparing the hearts of his students for *humanitas*." Medicine "is the science of helping, not of harming. Unless she strives in every way to succor the afflicted, she fails to provide for men the compassion [*misericordia*] that she promises" ([19], p. 24).

We see in Scribonius' statements two essential features of one who occupies the role of a physician: (1) He must be competent; and (2) he must be motivated by compassion and humaneness, that is, he must be a "lover of mankind" in the sense in which *philanthropia* and *humanitas* were popularly used at his time. Failure to act in a compassionate and humane manner rendered such a one no longer a physician. This is starkly different from the nearly negligible role that humanitarianism plays in the Hippocratic corpus. And it differs from Galen's thought in that Galen viewed competence as the only essential attribute of the physician, although the best physician was also a philosopher and such a physician should, as a consequence, also be a "lover of mankind". *Philanthropia* was, for Galen, highly desirable but not essential for a physician. Not so with Scribonius, for whom it was an essential feature of the true physician. So central were compassion and humaneness to Scribonius that he has been felt by some to have been 'nearly Christian'.[24] But he was not a Christian; he was a pagan significantly influenced by Stoic ideas that, at first glance, seem very similar to some Christian principles.

Philanthropia was a word not frequently used by early Christian writers.[25] Its occurrence is rare in the New Testament. The early Christians preferred a different word with a very different meaning: *agape*, a previously little-used and colorless word before it was given a specific Christian meaning.[26] The character of *agape* was rooted in the nature of God, "God is love [*agape*]," writes St. John (1 John 4: 8); hence *agape* was unlimited, freely given, sacrificial (because of God's love as revealed in the Incarnation), not dependent on the character of its object. Moreover, it was an active principle: love of God requires a love of mankind. "On these two commandments hang all the law and the prophets" (Matthew 22: 36–40). Membership in the Kingdom of God depends on the demonstration of active love that is manifested in charity and philanthropy (see Matthew 25: 31–46). One's love of God requires the spontaneous manifestation of love of one's brother; where there is no love of one's brother, there is no love of God (1 John 4: 20–21). Jesus said that his disciples would be known by their love for one another (John 13: 34–35). How is Christian love to be shown? Jesus gave the answer when asked, "And who is my neighbor?" The answer was contained in the parable of the good Samaritan: his advice was, "Go, and do the same" (Luke 10: 25–37).

It is not difficult to see the great gap that existed between the classical concept of *philanthropia* and the Christian idea of *agape* as the dynamic of ethics. Nor is it surprising that philanthropy has been called a peculiarly Christian product. Christian philanthropy had its roots in Judaism, in which the practice of love, mercy, and justice was a vital element in the worship of God (see Micah 6: 6–8). But the concept of *agape* led to a broadening and deepening of this philanthropic impulse. By the end of the second century, *philanthropia* begins to appear frequently in the Christian vocabulary, perhaps because it was a word (unlike *agape*) that pagans could readily understand. It is often used in the Fathers to describe God's love for man, for example, as shown in the Incarnation. By the fourth century it comes to be used as a synonym for *agape* in the liturgies of the Greek church.[27]

There has always been in the history of the Christian Church a tension, even if only latent, between secular medicine and theology, between the cure of the soul and the cure of the body. If indeed disease is in some way permitted by God, either for punishment or as a means of spiritual trial, then it is from God that we must expect healing to come. There have always been Christians who believed that the use of medicine is incompatible with the knowledge that it is God who heals. On the other hand, most Christians have seen no incompatibility with spiritual principles to assume that God can heal by means of physicians as well as without, and that physicians are ordained by God to relieve pain and sickness. If, however, suffering and disease are caused by a demonic spiritual force rather than by God, a physician treating by ordinary means would be of little help, and healing must be sought from God directly. A tension, between those who see medicine as an aspect of God's common grace and physicians as instruments of God, and those who hold that medicine is unnecessary since healing comes by faith and special grace, has characterized Christianity in every age.

There is little mention of Christians who were physicians in the early church. Aside from Luke, who is called "the beloved physician" by the Apostle Paul (Colossians 4: 14), we have no specific reference to any Christian physician before the mid-second century, when we hear of a Phrygian physician who was martyred at Lugdunum (Lyons) during the reign of Marcus Aurelius. It has been argued that a belief in miraculous healing was so prevalent in the early church that Christians saw little need for recourse to secular medicine.[28] According to this view, it took several centuries before Christians came to regard medicine as compatible with the life of faith. Anointing with oil, a practice for which there was apostolic injunction (see James 5: 14–15), seems to have been widely practiced in the early church for healing both by the clergy and laymen.[29] On the other hand, the fact that providential or miraculous healing was sought or claimed in the early church does not necessitate the assumption that Christians did not employ medicine or the services of physicians. When hospitals began to be established in the fourth century, as an outgrowth of Christian charity, it was common for them to employ physicians. Yet miraculous healing of disease continued to be sought and claimed, and because of the increasing tendency to seek help from relics, was perhaps more greatly emphasized than before.[30] It is not inconceivable that in the first centuries of the church Christians sought healing both through ordinary means and through attempts (such as unction and prayer) to obtain miraculous cures.

Because of the strong philanthropic motivation inherent in the Christian concept of *agape*, concern for others was shown not only in a desire for the salvation of souls through evangelism, but in providing help for those in physical need. Thus "pure religion" is defined in the Epistle of James as " . . . to visit the fatherless and widows in their affliction . . . . " (James 1: 27). In several passages in the Gospels Jesus is recorded as having set forth the principles of personal charity incumbent on all his followers. "For I was hungry and you gave me food, I was thirsty, and you gave me drink, I was a stranger, and you took me in, naked and you clothed me, I was sick, and you visited me, I was in prison, and you came to me . . . . Inasmuch as you have done it unto one of the least of these brethren of mine, you have done it unto me" (Matthew 25: 35–46). The verb used in this passage for visiting the sick often means "to care for, to succor," and it is sometimes used, in late classical Greek, to describe a physician's visiting a patient.

In the early church all believers were expected to visit and help the poor who were sick; this was especially true of women, who had time for this duty (Justin Martyr, *Epist. ad Zen. et Seren.*, 17; Tertullian, *ad Uxor.*, 2, 4). Churches set aside funds for the aid of widows and orphans, the sick and the poor, prisoners and travellers, and all those in need (Justin Martyr, *Apologia*, 1, 67). Deacons and deaconesses had a special duty to visit the sick and they reported cases of sickness or poverty to the local bishop (*Apostolic Constitutions*, 3, 19). They may have served also as nurses and attendants in hospitals, leprosaria, orphanages, and other charitable institutions maintained by the church ([6], p. 86).

Christians on a number of occasions showed heroic conduct in treating the sick during times of plague. One such occasion was during a plague at Alexandria during the reign of the Emperor Gallienus (260–268) when, following a period of persecution, Christians took an active part on behalf of those who were suffering from the pestilence. Their conduct is described in a letter written by Dionysius, bishop of Alexandria, which is preserved by Eusebius: "Heedless of the danger, they took charge of the sick, attending to their every need and ministering to them in Christ, and with them departed this life serenely happy; for they were infected by others with the disease, drawing on themselves the sickness of their neighbors and cheerfully accepting their pains" (Eusebius, *Ecclesiastical History*, 7, 22, 6–10).[31] Dionysius observes that a number of presbyters, deacons, and laymen lost their lives in the plague and were counted as equal to martyrs because of their piety and faith. The pagans, on the other hand, avoided their own sick relatives and even threw the sick into the streets before they were dead. The Christians were known for giving aid to Christian and pagan alike: the Roman Emperor Julian (361–363) complained that "the impious Galilaeans support not only their own poor but ours as well" ([50], p. 71).

In the fourth century, after persecution ended and Christianity gained the status of a legal religion, hospitals began to be established, together with orphanages and houses for the poor and for the aged. It was common practice to combine several such houses in one institution. One of the earliest and best known was the hospital or Basileias established by Basil the Great, who was bishop of Caesarea in Cappadocia. It was established about 372 ([6], p. 154). It included rooms for lepers and lodgings for travellers. The staff included nurses and medical attendants. Gregory of Nazianzus, who had seen the Basileias, describes it as "a new city, the treasure-house of godliness ... in which disease is investigated and sympathy proved .... We have no longer to look on the fearful and pitiable sight of men like corpses before death, with the greater part of their limbs dead, driven from cities, from dwellings, from public places, from water-courses. Basil it was more than any one who persuaded those who are men not to scorn men, nor to dishonour Christ the head of all by their inhumanity towards human beings"(*Oration* 20).[32] Hospitals (*xenodochia*) patterned after that of Basil spread throughout the Eastern Church and perhaps thence to the West. They were sometimes established by private persons, but more often by monastic communities or bishops. They were usually administered by bishops from church funds. Hospitals were recognized as particularly Christian institutions and the Emperor Julian urged, in a letter to Arsacius, the chief priest of Galatia, that pagan hospitals be established in every city for those in need, both for their own people and for foreigners (Julian, *Epistle* 49). The intention was to emulate the Christians.

In the early Middle Ages, during a period of great turbulence and confusion in Europe, monasteries became a refuge for the weak and defenseless, the sick, the poor, and the persecuted. The monastic clergy often took the lead in providing medical assistance through the establishment of *xenodochia* and hospices. There is ample evidence for a growing number of Christian physicians who were priests or monks already in late antiquity; after the fall of the Roman Empire in the West our sources reveal a strong emphasis on medical charity as an aspect of the monastic movement. In the *Rule of Saint Benedict* (sixth century) the cellarer is admonished to "take the greatest care of the sick, of children, of guests, and of the poor, knowing without doubt that he will have to render an account of all these on the Day of Judgement" (Benedict of Nursia, *Rule*, Chapter 31). Cassiodorus, who had been a Roman senator and chief minister to the Ostrogothic kings of Italy in the sixth century, at the age of seventy founded a monastery at Viviers. He was chiefly concerned to foster the pursuit of learning and the study of both classical and sacred literature, but he encouraged the monks to pursue medicine, bookbinding, and gardening. In his *Introduction to Divine and Human Readings*, Cassiodorus writes to those of his monks who were physicians, "I salute you, distinguished brothers, who with sedulous care look after the health of the human body and perform the functions of blessed piety for those who flee to the shrines of holy men – you who are sad at the sufferings of others, sorrowful for those who are in danger, grieved at the pain of those who are received, and always distressed with personal sorrow at the misfortunes of others, so that, as experience of your art teaches, you help the sick with genuine zeal; you will receive your reward from Him by whom eternal rewards may be paid for temporal acts" (Cassiodorus, *Institutiones*, 1, 31).

Henry Sigerist has written that Christianity introduced "the most revolutionary and decisive change in the attitude of society toward the sick. Christianity came into the world as the religion of healing, as the joyful Gospel of the Redeemer and of Redemption. It addressed itself to the disinherited, to the sick and afflicted, and promised them healing, a restoration both spiritual and physical .... It became the duty of the Christian to attend to the sick and poor of the community .... The social position of the sick man thus became fundamentally different from what it had been before. He assumed a preferential position which has been his ever since" ([36], pp. 69f). This verdict is, as we have seen, amply supported by the evidence. Compassion for the ill was a central feature of early Christianity that manifested itself in a manner that was foreign to classical medical ethics.

The physician as an ideal provided a commonplace for early Christian homiletics. In the Gospels Jesus appears as the Great Physician and the theme was much employed; he became the *verus medicus*, *solus medicus*, *verus archiater*, and was described as *ipse et medicus et medicamentum*, "himself both the physician and the medication."[33] The Church Fathers often drew on the practice of medicine for spiritual analogies. Their illustrations provide interesting insights into the ethical standards that were thought to be appropriate to physicians. Thus Origen attempted to imitate "the method of a philanthropic physician who seeks the sick so that he may bring relief to them and strengthen them" (Origen, *Contra Celsum*, 3, 74). The physician, he writes, manifests a Christ-like compassion in his care for the common man, the destitute, and the poor. Augustine speaks of the ideal physician who is motivated by charity and hence seeks no remuneration for his services but treats the most desperate cases among the poor with no thought of reward (Augustine, *Sermones*, 175, 8f). Because he is motivated by philanthropy the good physician is not deterred by fear of contagion.[34] Augustine believed that the physician should always be concerned for the cure of his patient (Augustine, *Sermones*, 9, 10), for if the physician were merely concerned about the practice of his art medicine would be cruelty (*idem*, *In Ps.*, 125, 14).

What influence such ideals had on the actual practice of medicine, we of course cannot say. There is evidence that many Christian physicians took seriously the philanthropic precepts of their faith. When Christianity displaced paganism in the fourth century and became an official religion, it became a religion of convenience for many, who perhaps found their principles of conduct and responsibility less tempered by their new faith. According to his biographer, Hypatios, who was a monk and a physician in the late fifth and early sixth centuries, treated patients afflicted with various sores who came to him because, being poor, they had been refused treatment by other physicians ([6], pp. 95f).[35] Apparently not all physicians (and we assume that those who were Hypatios' contemporaries were at least, for the most part, nominally Christian) took seriously the ideals of Christian medicine. The extent of a physician's conformity to the Christian ideal in medicine probably depended on his own Christian conviction and the nature of his faith. There were religious pressures against abortion and active euthanasia that may have deterred nominally Christian physicians from these practices. But there must have been many secular physicians of the period after the Christianization of the Empire who in other ways did not act very differently from their pagan counterparts of a few generations earlier.

There were those, however, and our sources indicate that they were not a few, who combined a commitment to their faith and to the medical art in such a way as to carry out the ideals of Christian philanthropy. Thus Augustine points to his friend, the physician Gennadius, a layman "of devout mind, kind and generous heart, and untiring compassion, as shown by his care of the poor" (Augustine, *Epistles* 159). Zenobius was a fourth-century priest and physician who is lauded by his biographer for serving his poor patients without remuneration and even helping them financially when necessary

([6], p. 182).[36] Basil wrote a letter to the secular physician Eustathius around 375, in which he extolls him for combining the medical and spiritual aspects of his art. "And your profession," he writes, "is the supply-vein of health. But, in your case especially, the science is ambidextrous, and you set for yourself higher standards of humanity, not limiting the benefit of your profession to bodily ills, but also contriving the correction of spiritual ills" (Basil of Caesarea, *Epistle* 189). It was in men such as these that the ideal of the Christian physician was recognized, an ideal in which spiritual and medical interests were blended into a concern for the care of the body *and* the soul.

The medical literature that has survived from the early Middle Ages is extensive and diverse and ranges from surveys of medical knowledge in the encyclopedic tradition to treatises dealing with specific areas of medical knowledge. Some extensive manuscripts are extant that contain many different treatises of the latter category and among these are occasionally found treatises that deal with medical etiquette and ethics.[37] While there is no way of knowing the exact purpose for which these treatises were written, they provide a valuable insight into medical ideals of the early Middle Ages. In one treatise the physician is exhorted to serve the rich and the poor alike, looking for eternal rather than material rewards. "Physician, care for the poor as well as the powerful. If the patient is rich you have a just occasion for profit; if poor, let one reward suffice." This "one reward", of course, is a spiritual one. The treatise ends with the exhortation, "Aid the sick, your reward coming from Christ, for whoever gives a cup of cold water in His name is assured of the eternal kingdom where with Father and Holy Spirit He lives and reigns for eternity." Another treatise states that the physician "should make the cases of others his own sorrow." The Christian physician must exercise charity: " . . . he should take care of the rich and poor, slave and free, equally for among all such people medicines are needed. Moreover, if certain compensation is offered, let him accept rather than refuse. If, however, it is not offered, do not demand it because, however much each one pays, the compensation for medical services cannot be equated with the benefits." The treatises from which these quotations are taken were all written before the tenth century and reflect ideals of monastic or at least clerical medicine.

There were secular physicians during the early Middle Ages but we know little about them and virtually nothing about their ethics. Somewhat before the twelfth century we begin to encounter secular physicians through their own writings. It is also about that time that two significant changes in the conditions of medical practice begin to occur. The first is the development of licensure requirements (whether imposed by external authority or obtained by medical guilds), which reflect a fundamental change in the very basis for the practice of medicine from a right to a privilege, with specific obligations attached to that privilege. The second is the attempt by moral theologians and casuists to define clearly the moral responsibilities of physicians. We shall briefly address the place of philanthropy in these two developments before we turn to the deontological writings of secular physicians of the late Middle Ages.

Throughout antiquity and the early Middle Ages there was no system of medical licensure. Anyone could call himself a physician and undertake to treat patients. The first effort in the Western World, datable with certainty, at limiting medical practice to those authorized by a governing authority occurred in 1140 in the Kingdom of Sicily under Roger II. The reason given for such restriction was concern for the safety and welfare of his subjects. Roger's legislation was strengthened by his grandson, Frederick II, in 1231, with the same justification. Frederick also imposed on physicians ca. 1241 the obligation to give medical advice to the poor without charge. The licensure policies of the Kingdom of Sicily were atypical of the time, however. Elsewhere in Europe licensure was being accomplished by guilds. While their early history is obscure, by the end of the twelfth century guilds were becoming common. One of the most striking features of late medieval urban life was its corporative aspect, particularly in its guild organization. Guilds (including university faculties or *collegia*) generally had the right to make and enforce standards of quality in their products or services, to control hours and working conditions, to limit competition between members, to limit entry into the craft or profession, and to ensure the proper treatment of customers. They held, or attempted to hold, a monopoly on their particular products or services. Part of their monopoly was the right to train and, essentially, to license new members. The claim was frequently made by the guilds that such restrictions were necessary to maintain a high degree of competence and ethics in the trade or profession. In the case of medical or surgical guilds, stress was continually placed on the dangers incurred by the general public when exposed to treatment by ill-qualified practitioners. Emphasis on the 'common good' appears frequently in the requests for the granting of guild charters or in pleas that existing rights to monopolies in medical or surgical services be protected. As a general rule it appears that medical or surgical guilds guaranteed to the community that they would do all things faithfully that pertain to their calling, i.e., dutifully serve the people, be available to attend those who need their services, charge reasonable fees, and police their own ranks. In some localities (e.g., in London and Montpellier) a medical or surgical guild would promise to treat the poor of the community gratis. Although such a practice was probably not at all uncommon, the extent to which the extending of free medical care to the poor was a practice of medical and surgical guilds in the late Middle Ages cannot be determined by the evidence available at present.

While the subject of medical charity is not frequently encountered in the available guild records, it was a matter of concern to individual physicians and church authorities. We shall first consider the latter. During the late Middle Ages, moral theologians and casuists directed considerable attention to defining the moral responsibilities of Christians generally, proscribing sins both of commission and of omission. They also addressed the moral responsibilities attached to those in various walks of life. Many of these sources, whose authors are usually referred to as summists, discuss the sins of physicians and surgeons. Much of this literature is in the form of guides to the confessor or confessional manuals designed to help the priest in his interrogation of the penitent. The sections devoted to physicians usually include discussions of expertise, diligence, and faithfulness to the traditions of the art, spiritual obligations, informing the patient of his condition, fees, and charity.[38]

In order better to understand the perspective of the summists whose statements we shall discuss shortly, a brief mention of the treatment of the obligation of medical charity by slightly earlier sources is warranted. There is a statement attributed to Pope Symmachus, quoted in the *Decretum* of Gratian (ca. 1140), to the effect that "there is not a great difference whether you inflict something fatal or allow it. He is proved to inflict death on the weak who does not prevent this when he is able to" (D. 83, 1. *Pars.*). This, of course, says nothing directly about physicians. Later, Joannes Teutonicus, in what became known as the *Glossa ordinaria* to the *Decretum* (ca. 1216 – 17), commented on this passage that the physician must treat both the poor and the rich gratis rather than allow them to die. This gloss is the *locus classicus* for the summists' discussion of the question whether a physician is obligated to cure gratuitously rather than to allow a sick person to die. Thomas Aquinas addressed the question of the extent to which the physician was morally bound to treat the poor gratuitously. Beginning with the comment that "no man is sufficient to bestow a work of mercy on all those who need it," he then suggested that one should first show kindness to those with whom one is united in some way. In regard to others, if a man "stands in such a need that it is not easy to see how he can be succored otherwise, then one is bound to bestow the work of mercy on him." Thus a lawyer is not always obligated to defend the destitute "or else he would have to put aside all other business and occupy himself entirely in defending the poor. The same holds with physicians in respect to attending the sick" (*Summa theologica*, 2–2, 71, 1). Astesanus (ca. 1317), the author of a confessional manual, followed Aquinas closely when discussing lawyers' duties to represent the poor gratuitously, adding that "the same must be said concerning a physician as to the care of paupers" (*Astesana*, 1, 39). The other summists tend to rely on Joannes Teutonicus' gloss on the above passage from the *Decretum*, considering the obligation especially to exist if the alternative to free care is the death of the patient. This distinction, however, is not seen in Antoninus's short *Confessionale* (1473), where he simply wrote that a physician has sinned "if he has not freely visited poor patients who he knew were not able to pay, because he is obligated to do that and even to pay for the medicine if he is able" (*Confessionale*, s.v., *Circa medicos*). This assertion does not express the limitations enunciated both by Aquinas and Joannes Teutonicus, and may well have sent the perplexed confessor to Antoninus's more extensive treatment in his *Summa* (1477). In his discussion of the nature of medicine, he said that the physician must treat paupers gratuitously if they are not able to pay, and not withdraw himself from their care "because this may be killing them indirectly" (*Summa theologica*, p. 277). Later, when treating the question in greater detail, he wrote that the physician "is not obligated to provide for all the poor ill simply and indiscriminately, but according to the place and time presenting itself . . . " (*Summa theologica*, p. 285). Generally the treatment by the summists[39] is along these lines with slight variations: a physician must give care and counsel freely to a sick pauper, "for he is proved to inflict death on the weak who does not prevent it when he is able to." In other words, he must give care and medicine without charge rather than allow a patient to die.

There is one matter that is conspicuously absent from these casuistic treatments of physicians' moral responsibilities and that is any reference to an obligation *to care*, i.e., any obligation to show personal concern. There are strongly-enunciated imperatives to extend medical care to the destitute and to treat the poor gratuitously, but there is no probing into that hinterland of the heart where genuine compassion and caring are found; there is no attempt to explore and discover the physician's motivation for practicing medicine, no forcing him to examine the 'why' of his practice, but only the 'how'. Had Christian charity – at least the medical charity expected of physicians – become merely the dead letter of the law, devoid of "the spirit that giveth life?"

While the church regarded medical charity as a duty incumbent upon physicians, individual physicians undoubtedly had mixed emotions in regard to such an obligation, at least insofar as we can judge from their writings. We shall next examine deontological treatises, composed primarily by secular physicians and surgeons. The deontological treatises of the late Middle Ages were written with the clear intention of providing other practitioners with, among other things, very practical and sometimes quite unethical advice on how best to survive in the profession. It is in the realm of fees, a subject with which the authors of this literature were quite preoccupied, that the most mercenary advice is found. Some, however, are quite idealistic. For example, in a short treatise ascribed to Constantine the African (d. 1087) is the statement that the physician "should not heal for the sake of gain, nor give more consideration to the wealthy than to the poor, to the noble than to the ignoble" ([30], p. 27). There is a medical treatise from a somewhat later period written by a member of the regular clergy (i.e., clergy living under a rule, e.g., monks, friars) for other regular clergy with the express intent that they might treat the poor gratuituously since the poor are often abandoned by ordinary physicians and surgeons. From the rich, however, these religious medical practitioners are to receive fees ([40], p. 96). The assertion that physicians desert those who cannot pay for their services is probably based on a suspicion of unchristian greed. The 'physicians' greed' is an extremely durable commonplace and medieval practitioners were quite sensitive to it, at least as revealed by their writings. That physicians were also unreligious was a common belief in the late Middle Ages and was often expressed by the adage *Tres medici, duo athei*. John Mirfeld, a cleric and amateur physician living in England in the late fourteenth and early fifteenth centuries, wrote that "the physician, if he should happen to be a good Christian (which rarely chances, for by their works they show themselves to be disciples, not of Christ, but of Avicenna and of Galen), ought to cure a Christian patient without making even the slightest charge if the man is poor; for the life of such a man ought to be of more value to the physician than his money" ([22], p. 132). Mirfeld then cites the same passage in the *Decretum* and the gloss thereon which the summists, as we have seen, also cite.

Physicians and surgeons in the late Middle Ages, if they were not regular clergy practicing medicine expressly for charity, were businessmen intent on making a living by their practice. Their advice on fees is quite varied. Perhaps the most crass is that from a southern Italian manuscript of the tenth century: "At the outset, accept at least half the remuneration without hesitation, for he who wishes to buy [your services] is disposed to pay and to beg [for treatment]. Get it while he is suffering, for when the pain ceases, your services also cease" ([30], pp. 23f). This advice is atypical and extreme. Guy de Chauliac (fourteenth century) writes that the physician should not be covetous or greedy, for if he is not he will receive a fee commensurate with his efforts, the financial ability of his patients, the success of the treatment, and his own dignity ([48], p. 357). John of Arderne (fourteenth century) suggests that the surgeon should "always be wary of asking too little, for ashing too little sets at nought both the market and the thing" ([48], p. 356). William of Salicet (thirteenth century) expresses the same opinion but adds that, for the sake of the reputation of the profession, the poor should be treated for nothing ([20], p. 157).

Henri de Mondeville (late thirteenth and early fourteenth centuries) nicely illustrates the quandary of the reasonably ethical physician of the late Middle Ages, who lived in a society where the imperative to charity was preached if not always followed and where societal expectations were often consonant with such an imperative. He advises the doctor not "to have too much faith in appearances. Rich people have a habit of appearing before him in old clothes, or if they do happen to be well dressed, they make up all sorts of excuses for demanding lower fees." Such people "claim that charity is a flower when they find someone else who will help the poor, and thus think that a surgeon should help the unfortunate; they, however, would never be bound by this rule .... I tell these people, then pay me for yourself and for three paupers and I will help them as well as you. But they never answer me, and I have never found a person in any position, whether clerk or layman, who was rich enough, or honest enough, to pay what he had promised until I make him do so." Mondeville advises that surgeons should be medical Robin Hoods: " ... the surgeon ought to charge the rich man as much as possible and get all he can out of them, provided that he does all that he can to cure the poor." His motivation for extending charity to the poor consisted of more than merely the advantages that might accrue to his reputation and to the honor of the profession, but was a product of enlightened self-interest, with eternal consequences, fully compatible with the theology of his time; "You, then, surgeons, if you operate conscientiously upon the rich for a sufficient fee and upon the poor for charity, you ought not to feat the ravages of fire, nor of rain nor of wind; you need not take holy orders or make pilgrimages nor undertake any work of that kind, because by your science you can save your souls alive, live without poverty, and die in your own house" ([20], p. 156; [48], pp. 356f).

During the late Middle Ages, Europe was periodically devastated by plague[40] beginning with the Black Death in 1348. The various strains of pestilential disease that arose were considered highly contagious, placing the conscientious physician in a delicate position. Public opinion impugned his actions as motivated by avarice if he appeared too willing to treat plague victims. If, however, he was unwilling to care for those ill with contagious disease he was open to accusations of irresponsibility or cowardice. Several chroniclers who lived at the time of the Black Death maintained that some physicians could not be enticed to treat the ill for any amount of money. Other physicians, however, attempted to care for plague victims without thought of remuneration. For example, one physician wrote in his diary that he had treated "out of compassion, as I would not have done it for money," a woman "who died of the worst and most contagious kind of plague, that of blood-spitting" ([1], p. 415).

Many physicians composed plague tractates, over 280 of which survive, in an effort to educate the public on prophylaxis and treatment. Many authors gave their reasons for writing their treatises, and for the most part they were composed *pro bono publico*. Johannes Jacobi wrote his in honor "of the Trinity and the Virgin Mary and for the utility of the republic and for the preservation of the healthy and the healing of the ill." Francischino de Collignano was moved "by pure love, affection and charity for all the citizens and especially for friends," and Michael Boeti wrote "in response to the requests of certain of my friends, for the service of God and for the common good." John of Burgundy concluded his tractate with this statement: " ... moved by piety and anguished by and feeling sorrow because of this calamity ... I have composed and compiled this work not for a price but for your prayers, so that when anyone recovers from the diseases discussed above, he will effectively pray for me to our Lord God .... " The author of an anonymous tractate wrote his, "sorrowing for the destruction of men and devoting myself to the common good and .... wishing health for all ... " ([1], pp. 418f).

Such motivation for the composition of medical literature and the dissemination of medical knowledge as that expressed in these tractates is not uncommon in medical history. As we have already seen, this was one manifestation of Galen's *philanthropia*. Those who write medical literature have one thing in common with all other authors: motivations as varied as the individuals themselves. Why does the poet compose his poetry, or the novelist his novel? For the love of art, as a response to a creative urge, for fame and glory, for money, for the benefit of mankind? All of these, and undoubtedly many others, can be attributed to artists as motivation for their work. When the last of these, the benefit of mankind, is the ostensible motive for the writing of medical literature (or for experimentation and research, for that matter), it is called philanthropy or beneficence, particularly if the author stands to make no money from the effort. If he receives remuneration for the work, however, particularly if the income from it is at least potentially great, his motivation is usually seen as tarnished if not entirely mercenary.

While the opportunity for substantial remuneration for writing was significantly increased by the invention of printing and later by copyright laws, in nearly all periods the potential (at least as popularly conceived) for a highly lucrative income was held out for medical and surgical practitioners.[41] Why does one pursue a medical career? One need merely look at the portrayal of physicians in much of Western literature, ancient, medieval, and modern, to find the answer: greed. But often at the periphery, and occasionally stepping onto center stage, is the noble physician, that selfless, caring, compassionate healer or consoler, the veritable epitome of altruism, philanthropy, and beneficence. In what stark contrast do these two types stand! What ambivalence is there in the layman's conception of the physician throughout the history of the Western world! Will Durant was undoubtedly right when he said that "in all civilized lands and times physicians have rivaled women for the distinction of being the most desirable and satirized of mankind" ([12], p. 531). This is evidenced by the existence of the most vitriolic denunciations of physicians alongside the most loving and admiring expressions of devotion to them throughout Western literature.

Henry Sigerist was quoted above as saying that "every period has an ideal physician in mind, indeed, must have one . . . . " Regardless of other variations in the ideal, two qualities seem always to be viewed as essential: competence and compassion. When competence is lacking, the would-be physician is seen as dangerous, however lofty his intentions; when compassion is lacking, he is regarded as coldhearted. When he lacks both qualities, he is denounced as a quack; but when he possesses both qualities, he is seen as an ideal physician. But the latter is not the only type properly called a physician. Very likely many people would agree with Galen's definition of a physician and for him the one essential feature was competence. Yet people may only refer to someone as a "true physician" who is, or at least approaches, an ideal. Edelstein, for example, in discussing Galen's ethics, says, " . . . the *true physician* [our italics] must become a philosopher himself, an adherent of Plato, or rather of the Platonism of Galen's time . . . " ([14], p. 335). It is rather unlikely that Galen would have agreed with Edelstein's assessment of his position, since Galen clearly considered competence as that which distinguished a *true* physician from one who was not really a physician at all. For Galen, the *best* physician also had to be a philosopher. When summarizing Scribonius' views, Edelstein uses the same expression, writing that Scribonius regarded as a *true physician* one who harms no one and possesses the qualities that he considered as essential to the *officia* of one who occupies the *professio* of a physician. Edelstein is quite correct in his appraisal of Scribonius' attitudes since for the latter *humanitas* was a *sine qua non* for being a physician at all. In discussing Asclepius, the Greek god of medicine, and the apolitical nature of his cult, Edelstein writes, "As a *true physician* [our italics], Asclepius was interested in individuals rather than in politics. His only concern was the disease and the patient who came to him in quest of help" ([13], p. 180). This is probably a statement of Edelstein's own attitude of what a physician ought to be. It does not reflect historical values necessarily typical of antiquity and clashes with Plato's ideal physician who would be very political in his treatment, concerned with eugenics and the welfare of the state over the welfare of the patient.[42]

Who is a 'true physician'? Temkin writes that "different people have chosen the profession of medicine for different reasons, some lofty and some not so lofty. Respect for life is a necessary condition for being a physician; it is not a necessary motive for entering medicine. The question of how deeply respect for life must be felt, as long as responsibility and duty are not shirked, must be left open."[43] We said above that many people would regard competence as the one essential quality of a physician; indeed it is probably safe to say that most would, until faced with extreme instances, for example, of the competent but evil physicians who performed their diabolical experiments in the Nazi deathcamps. Few would say that these were 'true physicians' at all, while many feel that the performance of a score or more abortions each day by some physicians should hardly disqualify them from being called 'true physicians'.

There is something terribly elusive about the qualities of a 'true physician' and there will probably never be unanimity in any definition advanced. During the eras discussed in the present paper, was that nebulous quality, partially expressed by the words caring, compassion, humanitarianism, altruism, beneficence, or philanthropy, considered an essential feature of a 'true physician'? To that question an unequivocal yes or no simply cannot be given. We can safely say that some would answer the question with a hearty "yes". Surely Scribonius would, as would those, at any period, who have felt deeply the imperative of Christian *agape*. It is much more difficult when we look for those who, upon probing reflection, would respond with an unequivocal negative. Well might they stress that competence is the only essential positive feature; caring, compassion, and so forth, although desirable, are not essential to a 'true physician', although the presence of the opposites of these qualities would disqualify a candidate for such a designation. Surely the *actively* uncaring, uncompassionate physician would be felt by most, if not all, at any time, in any place, to be violating something inherent in the calling of the physician.

Even the most idealistic, however, would have to concede that if a 'true physician' must be caring, compassionate, humanitarian, altruistic, beneficent, and philanthropic, many or probably most physicians at any time simply would not be 'true physicians'. For both the realist and the idealist usually recognize that such qualities cannot meaningfully be required or imposed on one who does not wish to possess them, or does not possess them naturally. They are too internal, too much a matter of the heart to be *enforceable* criteria for admission to any walk of life. Witness the history of the clergy, those in whom such qualities typically are even more expected than they are in physicians. Is the uncaring pastor or priest truly a clergyman at all? Will it not depend entirely on how the office, and the field as well, are viewed? If any field is seen to be mechanistic and its practice mechanical, the answer will likely be "yes". If expectations (should we say standards?) are higher, the answer will almost undoubtedly be "no". But the question will remain a subject of debate probably for all professions in which those nebulous qualities are strongly viewed by many, both within and outside the professions, as being very highly desirable if not indispensable.

Western Washington University, Bellingham, Washington
Oregon State University, Corvallis, Oregon

NOTES

[1] *Precepts* is a relatively late work. Ludwig Edelstein maintains that it "cannot have been written before the first century B.C. or A.D." ([14], p. 322).

[2] In *The Old Humanities and the New Science*, as quoted by Edelstein in [14], pp. 319f.

[3] On *philanthropia*, see [16], pp. 102–117; and [8].

[4] See [21], pp. 77–88; and [11].

[5] The exact antonym for *philanthropia* is *misanthropia*, "hatred of men". *Apanthropia* is "inhumanity".

[6] Edelstein ([14], p. 322) dates this treatise to 350–300 B.C. at the earliest.

[7] For a convenient collection of translations of several such decrees, see [21], pp. 202ff.

[8] See [41], p. 110.

[9] E.g., Thucydides, *The Peloponnesian War*, 6, 14; Euripides, *The Phoenician Women*, 893; Plato, *The Statesman*, 293 A–C; Laws, 862 B, 720 D–E (cf. *Gorgias*, 464 B); *Republic*, 340 C ff; Aristotle, *Nicomachean Ethics*, 1180 b; *Politics*, 1287 a; Pseudo-Demosthenes, *Against Aristogeiton*, 2, 26; Aeschines, *Against Ctesiphon*, 225f. (cp. Cicero, *Republic*, 1, 62; 5, 5); Cicero, *De oratore*, 2, 186; *Disputations*, 3, 82. See also Livy, 42, 40, 3.

[10] For a discussion, see [32], Chapter 2., esp. pp. 27 and 38.

[11] This work is usually cited by its abbreviated Latin title, *De placitis*. The section under discussion is 9, 5, in [19], Vol. 5, pp. 751f. A critical text and English translation by Phillip LeLacy is available as Vol. 5, 4, 1, 2, of the *Corpus Medicorum Graecorum*.

[12] Under the Roman Empire some physicians were granted exemption from certain burdensome duties. See [33].

[13] Menodotus, who lived in the late first, early second centuries A.D., was leader of the empirical school of medicine.

[14] Diocles of Carystus, who lived in the fourth century B.C., was a physician known for his scientific originality.

[15] Empedocles, fifth century B.C., although better known as a philosopher, was of some reputation as a physician. Galen considered him to be the founder of the Sicilian medical school.

[16] See [41], pp. 79–105.

[17] An early and remarkable example of this feeling, perhaps from the Cynic point of view, is a poem written by Cercidas of Megalopolis (third century B.C.?), a politician who urges the upper classes to avert a social revolution by caring for the sick and giving to the needy. His attitude is genuinely humanitarian and not merely self-serving. For a translation of the poem, see [3], pp. 58–59. See also p. 52.

[18] See [4], pp. 31–37.

[19] See [5], pp. 135–145; [39], pp. 246–263; and [16], pp. 115–117.

[20] That Galen was independently well-to-do owing to his inheritance and the vast honoraria received from his wealthy patients undoubtedly made his *philanthropia* easier. See [42], p. 47; and [27], p. 453.

[21] *τέχνη οὕτω φιλάνθρωπος.*

[22] The main reason behind the refusal to be a physician to the Persian satrap is given in the pseudo-Hippocratic "letter", No. 5 in [28], Vol. 9, pp. 400ff: He, as a Greek, will not treat barbarians who are enemies of the Greeks. See [26], pp. 94f.

[23] Published separately by Karl Deichgräber in [9].

[24] See [26], pp. 95f; and [14], pp. 339, 344 and n. 45. Space does not permit our discussing two other interesting documents that express attitudes harmonious with Scribonius': (1) An inscription of the late second century A. D., a poem entitled *On the Eternal Duties of the Physician*: see [29]. (2) A *progymnasma* (i.e., rhetorical exercise) of Libanius (fourth century A.D.) dealing with medical deontology. No English translation seems to exist. The Greek text is in [17].

[25] See [25], Vol. 9, *s.v.*, *φιλανθρωπία*, esp. pp. 111–112; and [16], pp. 111ff.

[26] See [25], Vol. 1, *s.v.*, *ἀγαπάω*, pp. 21–55; and [16], pp. 227–243.

[27] See [10], esp. pp. 204ff.

[28] See e.g., [7], [18], and [24].

[29] See [38], Vol. 2, *s.v.*, "Unction", p. 2004.

[30] See [47], pp. 35–69.
[31] The translation quoted is found in [49], p. 305.
[32] Quoted in [38], Vol. 1, *s.v.*, "Hospitals", p. 786.
[33] There is a fairly extensive literature on the motif of *Christus Medicus*. See, e.g., [34], [2], and [35].
[34] See Basil of Caesarea, *Epistles* 189, written to the physician Eustathius.
[35] Ambrose, a century and a half earlier, had commented that physicians left the care of lower-class patients to their slaves or servants (*medicorum pueri* and *ministri*). "Let the rich man," he writes, "call the master, the poor man the servant" (*Enarratio in Psalmum*, 36, pr. 3).
[36] Constantelos also mentions the physician Sampson (sixth or perhaps fifth century) who "transformed his home into a free public clinic. Not only did he treat poor patients free of charge, but also offered them food and lodging" ([6], p. 182).
[37] For a translation of many of these, see [30].
[38] The following works in the genre of confessional literature were used in preparation of this section:

Artesanus de Asti, *Summa de casibus conscientiae* (c. 1317; Venice, 1478), copy at Free Library of Philadelphia; generally cited as *Artesana*.

Bartholomaeus de Sancto Concordio, *Summa casuum* (c. 1338; Venice, 1473), copy at University of Pennsylvania; generally cited as *Pisanella*.

Antoninus of Florence, *Confessionale-defecerunt* (1473; Esslingen, 1474[?]), copy at College of Physicians of Philadelphia.

Antoninus of Florence, *Summa theologica* (or *Summa moralis*) (1477; 1740; reprint, Graz: Akademische Druck- und Verlagsanstalt, 1959).

Bartholomaeus de Chaimis, *Interrogatorium sive confessionale* (c. 1474; Nuremberg, n.d.), copy at Free Library of Philadelphia.

Baptista Trovamala de Salis, *Summa de casibus conscientiae* (c. 1480; Venice, 1495), copy at College of Physicians of Philadelphia; generally cited as *Baptistina*.

Angelus Carletus de Clavasio, *Summa Angelica de casibus conscientiae* (c. 1486; Lyons 1494), copy at Free Library of Philadelphia; generally cited as *Angelica*.

Cajetan (Tommaso de Vio), *Summula peccatorum* (1525; Florence, 1525), copy at University of Pennsylvania.

Bartholomaeus Fumus, *Summa Armilla* (c. 1538; Cologne, 1627), copy at Catholic University of America.

[39] Pisanella, *s.v.*, *Medicus vel cirurgicus*; Antoninus, *Summa theologica*, pp. 284f, *Baptistina*, *s.v.*, *Medicus vel cirugicus*, 7; *Angelica*, *s.v.*, *Medicus*, 5 and 6; Chaimis, *Interrogatorium*, *s.v.*, *Medicis, phisicis, et cirogicis*; Fumus, *Summa Armilla*, *s.v.*, *De Medico*, 1 and 2.
[40] On the ethical standards of physicians during plague, see [1].
[41] Physicians' wealth, or at least potential for wealth, is a pervasive motif in many cultures. See, e.g., [44], *s.v.*, "Physician".
[42] See, e.g., *Republic*, 407 C–D.
[43] "Respect for Life in the History of Medicine," in [43], p. 18.

BIBLIOGRAPHY

1. Amundsen, D. W.: 1977, 'Medical Deontology and Pestilential Disease in the Late Middle Ages', *Journal of the History of Medicine and Allied Sciences* **32**, 403–421.
2. Arbesmann, R.: 1954, 'The Concept of "Christus Medicus" in St. Augustine', *Traditio* **10**, 1–28.
3. Barker, E.: 1956, *From Alexander to Constantine*, Clarendon Press, Oxford.
4. Bell, H. I.: 1948, 'Philanthropy in the Papyri of the Roman Period', *Hommages à Joseph Bidez et à Franz Cumont*, Collection Latomus II, Brussels, pp. 31–37.
5. Clarke, M. L.: 1968, *The Roman Mind*, Norton, New York.
6. Constantelos, D. J.: 1968, *Byzantine Philanthropy and Social Welfare*, Rutgers University Press, New Brunswick, N. J.
7. Dawe, V. G.: 1955, *The Attitude of the Ancient Church Toward Sickness and Healing*, unpublished doctoral dissertation, Boston University School of Theology, Boston.
8. Déaut, R. le: 1964, 'Philanthropia dans la Littérature grecque Jusqu'au Nouveau Testament', *Studi e Testi: Mélanges Eugène Tisserant* **1**, 255–294.
9. Deichgräber, K.: 1950, *Professio Medici, Zum Vorwort des Scribonius Largus*, Abhandlungen der Akademie der Wissenschaften und der Literatur, No. 9.
10. Downey, G.: 1955, 'Philanthropia in Religion and Statecraft in the Fourth Century After Christ', *Historia* **4**, 199–208.
11. Downey, G.: 1965, 'Who is My Neighbor? The Greek and Roman Answer', *Anglican Theological Review* **47**, 2–15.
12. Durant, W.: 1953, *The Story of Civilization*, Vol. 5, Simon and Schuster, N.Y.
13. Edelstein, E. J. and Edelstein, L.: 1945, *Asclepius: A Collection and Interpretation of the Testimonies*, Vol. 2, The Johns Hopkins Press, Baltimore.
14. Edelstein, L.: 1967, in O. Temkin and C. L. Temkin (eds.), *Ancient Medicine: Selected Papers of Ludwig Edelstein*, The Johns Hopkins Press, Baltimore.
15. Entralgo, P. L.: 1969, *Doctor and Patient*, trans. by F. Patridge, McGraw-Hill, New York.
16. Ferguson, J.: 1958, *Moral Values in the Ancient World*, Methuen, London.
17. Foerster, R. (ed.): 1915, *Libanii Opera*, Vol. 8, Teubner, Leipzig, pp. 182–194.
18. Frost, E.: 1949, *Christian Healing*, Mowbray, London.
19. Galen: 1823, in C. G. Kühn (ed.), *Opera Omnia*, (rpt. 1965), George Olms, Hildesheim.
20. Hammond, E. A.: 1960, 'Income of Medieval English Doctors', *Journal of the History of Medicine and Allied Sciences* **15**, 154–169.
21. Hands, A. R.: 1968, *Charities and Social Aid in Greece and Rome*, Cornell University Press, Ithaca.
22. Hartley, P. H. S. and Aldridge, H. R.: 1936, *Johannes de Mirfeld of St. Bartholomew's, Smithfield: His Life and Works*, Cambridge University Press, London.
23. Jones, W. H. S. (trans.): 1923–1931, *Hippocrates*, Loeb Classical Library, Harvard University Press, Cambridge.
24. Kelsey, M. T.: 1973, *Healing and Christianity in Ancient Thought and Modern Times*, Harper and Row, N.Y.
25. Kittel, G. and Friedrich, G. (eds.): 1964–1976, *Theological Dictionary of the New Testament*, trans. by G. W. Bromiley, Eerdmans, Grand Rapids.

26. Kudlien, F.: 1970, 'Medical Ethics and Popular Ethics in Greece and Rome', *Clio Medica* 5, 91–121.
27. Kudlien, F.: 1976, 'Medicine as a "Liberal Art" and the Question of the Physician's Income', *Journal of the History of Medicine and Allied Sciences* **31**, 448–459.
28. Littré, E.: 1839–1861, *Oeuvres Complètes d' Hipprocrate*, J. B. Bailliere, Paris.
29. Maas, P. L. and Oliver, J. H.: 1939, 'An Ancient Poem on the Duties of a Physician', *Bulletin of the History of Medicine* 7, 315–323.
30 MacKinney, L. C.: 1952, 'Medical Ethics and Etiquette in the Early Middle Ages: The Persistence of Hippocratic Ideals', *Bulletin of the History of Medicine* **26**, 1–31.
31. May, W. F.: 1975, 'Code, Covenant, or Philanthropy', *The Hastings Center Report* **5**, 29–38.
32. McNeill, J. T.: 1951, *A History of the Cure of Souls*, Harper and Row, N. Y.
33. Nutton, V.: 1971, 'Two Notes on Immunities: *Digest* **27**, 1, 6, 10, and 11', *Journal of Roman Studies* **61**, 52–63.
34. Pease, A. S.: 1914, 'Medical Allusions in the Works of St. Jerome', *Harvard Studies in Classical Philology* **25**, 73–86.
35. Schipperges, H.: 1965, 'Zur Tradition des "Christus Medicus" im frühen Christentum und in der älteren Heilkunde', *Arzt und Christ* **2**, 12–20.
36. Sigerist, H.: 1943, *Civilization and Disease*, Cornell University Press, Ithaca.
37. Sigerist, H.: 1961, *A History of Medicine*, Vol. 2, Oxford University Press, N. Y.
38. Smith, W. and Cheetham, S.: 1875–1880, *A Dictionary of Christian Antiquities*, John Murray, London.
39. Snell, B.: 1960, *The Discovery of the Mind*, trans. by T. G. Rosenmeyer, Harper and Row, N. Y.
40. Talbot, C. H.: 1967, *Medicine in Medieval England*, Oldbourne, London.
41. Tarn, Sir Wm. and Griffith, G. T.: 1952, *Hellenistic Civilization*, 3rd ed., Arnold, London.
42. Temkin, O.: 1973, *Galenism: Rise and Decline of a Medical Philosophy*, Cornell University Press, Ithaca.
43. Temkin, O., *et al.*: 1976, *Respect for Life in Medicine, Philosophy and the Law*, The Johns Hopkins Press, Baltimore.
44. Thompson, S.: 1955–1958, *Motif-Index of Folk Literature*, University of Indiana Press, Bloomington.
45. Veatch, R. M.: 1979, 'Professional Medical Ethics: The Grounding of Its Principles', *The Journal of Medicine and Philosophy* **4**, 1–19.
46. Walzer, R.: 1949, 'New Light on Galen's Moral Philosophy (from a recently discovered Arabic source)', *Classical Quarterly* **43**, 82–96.
47. Warfield, B. B.: 1918 (rpt. 1972), *Counterfeit Miracles*, The Banner of Truth, London.
48. Welborn, M. C.: 1938, 'The Long Tradition: A Study in Fourteenth Century Medical Deontology', in J. L. Cate and E. N. Anderson (eds.), *Medieval and Historiographical Essays in Honor of James Westfall Thompson*. University of Chicago, Chicago.
49. Williamson, G. A. (trans.): 1975, *Eusebius: The History of the Church from Christ to Constantine*, Augsburg, Minneapolis.
50. Wright, W. C. (trans.): 1913, Julian, *Letters*, Vol. 3, Loeb Classical Library, London.

ALLEN E. BUCHANAN

PHILOSOPHICAL FOUNDATIONS OF BENEFICENCE

The place of beneficence in health care may seem so central and obvious as to raise no serious issues for philosophical reflection. After all, the role of the physician and of other health care providers is usually defined as a helping role. And insofar as the health care provider's role involves not only taking care of, but also caring, his beneficent behavior is supposed to be the outward expression of an inward concern. On this view the provider's beneficence – action intended to do good – is, or should be, motivated by benevolence – a direct regard for the good of others. Indeed, if it is assumed that health is a preeminent good for human beings, then conduct directed toward securing this good is one of the highest forms of beneficence, and where it is an expression of a benevolent attitude, that attitude would seem to deserve the highest moral praise.

Behind these apparent truisms lies a tangle of perplexing issues. Some argue that if beneficence is essential to the provider-patient relationship, then a profit-oriented, fee-for-service system is singularly inappropriate, since it often pits the patient's good against the provider's prosperity and nurtures self-regard at the expense of benevolence. Some even suggest that since the patient-provider relationship is properly understood as a beneficent relationship founded on trust, attempts to construe it as a contract, an exchange agreement among equals for mutual benefit, is misguided. Others contend that since health care is like any other important service, there is no more reason for the health care provider to forswear profit as his primary goal than for those in other occupations to do so, and that the quality of care need not suffer in consequence, so long as there is a free, competitive market for it.

Regardless of which model of the provider-patient relationship is assumed, and independently of the problem of conflicts of interest between provider and patient, the provider's duty to do what is best for one patient may conflict with the same duty toward others, since health care resources, including the provider's time, will often be scarce. Further, even if we set aside both the problem of conflict of interest between provider and patient and the conflict of interest among patients due to scarcity of resources, there remains the exceedingly complex problem – partly conceptual, partly moral, and partly

Earl E. Shelp (ed.), Beneficence and Health Care, 33–62.

empirical – of correctly identifying the individual's good. This latter problem is especially acute where the patient is terminally ill or so reduced in his capacity for cognition or even for sensation that the very notion of *his interests* becomes attenuated.[1]

Finally, even if we abstract from all interpersonal conflicts of interest and assume that we have properly identified the patient's good, the platitude that the provider's overriding duty is to act for the good of the patient is itself dubious. For it appears that an exclusive concern for the patient's good or interests ignores his rights and expresses a paternalistic conception of the provider-patient relationship. Granted that autonomy and beneficence may conflict, what limits does respect for a person as an autonomous agent impose on our efforts to do good? Are there moral limits on what we may do to achieve what a person recognizes as his own good but is unable to achieve for himself?

At the level of institutional policy all semblance of consensus about the proper role of beneficence evaporates just as quickly. Institutional schemes for rationing scarce health care resources may avoid a situation in which the individual provider must choose to bestow his services on some known individuals rather than on others; but someone must make the decision at a higher level. Perhaps the most fundamental disagreement is over the question of whether certain forms of beneficence, including the provision of some types or levels of health care, are required by *justice* or are properly classified as acts of *charity* or *generosity*. The answer to this question has momentous implications for the role of government in the provision of health care and other important goods, assuming that the requirements of justice may, if necessary, be enforced by the coercive power of the state.

Once the urgency of these issues is recognized, two more fundamental queries become inescapable. First, what are the scope and limits of beneficence in health care? Who are the proper recipients of our efforts to do good: only those who stand to us in certain special relationships, only those who are members of our own society, mankind more generally, or at least the most needy thereof? Or should the recipients of our beneficence include future generations as well (since they will surely need some of the resources now at our disposal)? Second, how useful is a conception of the proper role of beneficence as a fundamental tool for the reasoned moral assessment of individual relationships and institutional structures in health care? In other words, to what extent can such a normative conception of beneficence serve as a basis for judgments about what is praiseworthy and what is deficient in health care as it exists today, and as a guide for change?

Neither of these questions can be adequately answered until a more general philosophical theory of beneficence is available. Such a theory would include at least the following elements: (a) a preliminary, clarifying analysis of the concept of beneficence, (b) a specification of the relationship between beneficence and justice; and (c) an account of the scope and limits of the moral requirements of beneficence that addresses both the problem of conflicts of interest among possible recipients of beneficence and of conflicts between beneficence and respect for autonomy. Further, it appears that none of these elements can be adequately developed, nor systematically related to the others, until an account of the relationship between rationality and beneficence is explored. The aim of this essay is not to provide such a theory or even to apply such a theory to ethical issues in health care. My goal, rather, is to begin the task of ascertaining what an adequate theory of beneficence would look like and to do so in a way that emphasizes the dependence of the other elements of the theory upon an account of reasons for beneficent action. I shall frequently draw examples from health care, but my concern is only to prepare the ground for investigations of beneficence in health care.

While it would be naive to expect that an analysis alone of the meaning of the term 'beneficence' will itself answer the moral questions posed above, some preliminary distinctions will prove helpful. First, as was noted earlier, we may distinguish beneficence, as action intended to do good to another, from benevolence, as a motive or practical attitude of direct concern for the good of others. I may act with the intention of doing good to Jones, even if my motive in doing so is a direct concern for my own good, with only an indirect concern for Jones, derivative on my belief that benefiting Jones is a mere means toward benefiting myself. Second, even where my motive in acting beneficently toward Jones is direct in the sense of not being derivative in this instrumental way on some other motive, such as self-interest, there are still at least two significant cases to distinguish. On the one hand, my direct concern for Jones's good may issue from my judgment that I ought morally to help Jones; on the other hand, it may be simply an unreflective desire that Jones's good be achieved.

These are not idle distinctions; they are relevant to some of the more important issues raised earlier. In particular, they bear directly on the problem of specifying the scope and limits of beneficence and on the task of determining the role which alternative institutional arrangements can play in either facilitating or hindering the promotion of the good of others. For example, if a moral theory is to provide guidance concerning the proper scope and limits of beneficence, it must operate within a set of realistic assumptions about the practical efficacy of the motive of benevolence as compared with self-regarding motives, and it must enable us to determine to what extent the morally virtuous person not only acts beneficently, but does so out of direct regard for the good of others. Similarly, once this distinction between action that promotes another's good and action intended to do so, and the distinction between promoting another's good out of direct concern for his good and promoting another's good out of self-interest are admitted, it becomes an empirical question whether the good of others is best promoted through beneficence or through certain forms of mutually beneficial but self-interested behavior, such as market exchanges. And once this is seen to be an empirical question we are also confronted with the issue of whether various interpersonal and institutional problems in health care are properly diagnosed as resulting from a lack of benevolence, or improperly directed beneficence, or a failure to appreciate limitations on the effectiveness of beneficence.

The task of discovering or constructing rational foundations for beneficence may be approached *via* a taxonomy which distinguishes beneficence *in special* relationships from *generalized beneficence*. Special relationships involving beneficence may be either contractual or non-contractual. Examples of contractual beneficence relationships include ordinary exchange agreements for mutual benefit: Jones agrees to help Smith harvest his crop on the condition that Smith will do the same for him. Among the most important special relationships involving beneficence that are usually said to be non-contractual are friendship, love, and familial relationships.

It is characteristic of contractual special relationships that, so far as the relationship is viewed simply as contractual, it is assumed that there is no direct motive for promoting the other party's good. Instead, the contractual agreement is seen as an artifice to achieve at least a temporary coincidence of interests. The terms of the agreement are designed to assure each party that his contribution to the good of the other will be reciprocated.

Moral philosophers have often noted that in intimate relationships, beneficence is not to be understood according to a contractual model because that model assumes that the interests of the members of the relationship are not only distinct but potentially, if not actually, in conflict. In contrast, the members of an intimate relationship are seen as elements of an affectively integrated whole, who promote each other's good spontaneously out of love or direct concern, rather than from instrumental calculations of self-interest. According to this way of thinking, a 'friendship' which needs an artifice to produce a coincidence of interests by specifying terms which explicitly link the interests of one party with the distinct and potentially conflicting interests of the other, and which requires a mechanism for assuring each party that either the terms of the agreement will be kept or some form of damages or compensation will be paid, would not be a genuine friendship.

Granted this way of contrasting intimate special relationships with contractual special relationships, it is not surprising that moral philosophers have tended to concentrate on providing justifications for beneficence in the latter rather than in the former. Reasons for a parent's beneficence toward her child are often given, but they are usually reasons of a singularly uninformative sort. Setting aside theological justifications (e.g., "God commands parents to care for their children"), the most natural reply to a request for a reason why a parent should care for her child may be one which merely emphasizes that a special relationship exists ("After all, this *is* my child"). Similarly, in most contexts the appropriate reply to a request for the reason one ought to help a friend will be an assertion that a special relationship exists, not a further reason over and above this fact.

It may be said, however, that even if it is unusual for parents to articulate their reasons for promoting the interests of their child, there *is* a reason – namely, that it is they who have brought this being, vulnerable and dependent as it is, into the world. Similarly, even if it is unseemly for me to hanker after reasons why I should help my friend in his hour of need, reflection on the nature of the relationship may reveal reasons. Since in my interactions with this person I have, both implicitly and explicitly, presented myself as his friend, I have communicated to him certain feelings of special regard. For this reason, a special relationship such as friendship may generate special reasons for beneficence; as with the parent-child relationship, it involves patterns of conduct which nurture expectations of beneficent actions. This is not to say, of course, that all special relationships in which one person has helped another generate legitimate expectations of future aid. This is not so in the case of most contractual relationships, since the terms of the contract are designed not only to fix but also to limit expectations of benefit. Further, unless the past benefits were freely received, rather than imposed, and unless the relationship itself was freely entered into by both parties, past beneficent behavior may generate no legitimate expectations of future beneficence, and therefore the existence of such expectations may not serve as a reason for acting beneficently.

There is another class of special relationships involving beneficence which shares some but not all features of contractual special relationships. These special relationships, which may be called quasi-contractual, include various forms of promissory relationships other than contracts, where the latter are understood as promises of exchange for *mutual* benefit. I may promise to help you move your furniture without specifying any terms for reciprocation on your part. Yet so far as we regard such a promise as generating a special *right* on your part to my performance of what I promised, beneficence based on promising seems closer to the model of contractual beneficence than to that of intimate relationships. This is so because talk about rights is characteristically appropriate only where there is at least a serious potential for conflict of interest between the right-holder and the one against whom he may make a claim of right. In the case of promising, talk about a special right of the one to whom the promise is made is useful only if it is assumed that there may be a divergence between what was promised and what the promiser would otherwise be inclined to do.

Requests for a rational foundation or justification for beneficent action in those special relationships which are contractual or quasi-contractual may be met initially with the same sort of reply we encountered earlier in the case of special non-contractual relationships: e.g., "Why ought I to help Jones? – because I promised I would." And in the usual sort of circumstances this, too, will be accepted as sufficient; no further reason, beyond the mere fact that one promised, will be required.

Nonetheless, it is possible to push the question of justification deeper, asking why it is that the fact that one promised is a reason for doing what one promised, just as it is possible to ask why the fact that the other party to a contract has performed his part of the bargain is a reason to perform one's own. One obvious response noted earlier applies to contracts as well as promising. In both cases expectations are created and it is plausible to contend that it is the existence of these expectations, or rather the importance of not thwarting them, which supplies a reason for performance. Though this response is more informative than a mere repetition of the statement that one did after all promise, or that the other party did perform, it need not be a conversation-stopper. For even if – due to your abysmal record as a promise-keeper – I have no expectation that you will keep your promise, but at most a lingering hope based on my groundless faith in the possibility of moral regeneration, this does not mean that there is no reason for you to keep it.

Reflections along these lines have led moral philosophers to seek an even deeper justification for keeping promises and contractual agreements, including those where the action in question is to promote the good of another. It has been argued that the fact that you promised can serve as an adequate reason for your doing what you promised only because our particular interaction falls under a social practice which itself is rationally justified.[2] The reason for promoting another's good in a contractual or promising relationship is then seen to be derivative on the reasons for participating in and supporting the practice or institution which makes such relationships possible.

This strategy of displacing the task of justification from the level of particular acts to the level of social practices or institutions pushes us further toward attempts to develop systematic moral theories which aim to provide rational foundations for the whole range of our commonsense moral judgments concerning beneficence. Though, as we saw earlier, there seems to be nothing incoherent about the request for deeper levels of justification in the case of non-contractual relationships, moral philosophers have tended to concentrate on the quest for rational foundations for beneficence in contractual or quasi-contractual relationships and for generalized beneficence, rather than for beneficence in intimate relationships, for several reasons. First, there is the traditional idea that familial relationships, friendships, and loving relationships are *natural* forms of human interaction, not social artifices, like promising or contracts, and that consequently it is inappropriate to ask for a justification for them (though it may not be out of order to seek an evolutionary or functional explanation of them). The immediate difficulty with this response is that it overlooks the fact that even the most intimate relationships may have a large conventional component and that there is no longer a consensus on whether the nuclear family, monogamous sexual relationships, or some traditional forms of friendship are natural, rational, or even moral. Second, it may have been erroneously assumed that because members of a harmonious family, a sincere friendship, or a flourishing loving relationship do not themselves find it necessary or appropriate to justify their beneficence, the quest for philosophical foundations for beneficence in such relationships is misguided. Third, the need for justifications seems less acute in the case of intimate relationships than in generalized beneficence or in contractual or quasi-contractual relationships if the problem of justification is assumed to be that of providing reasons for a person to promote interests that are wholly distinct from and frequently in opposition to his own. But while it may be true that the presence of conflict of interest makes the task of justification more onerous in the case of general beneficence or beneficence in contractual or quasi-contractual special relationships, we shall see that attempts to provide systematic justifications for beneficence in the latter cases can have important implications for beneficence in intimate relationships as well. Nonetheless, the main concern of this essay is with generalized beneficence, and to a lesser extent on beneficence in impersonal special relationships.

Commonsense morality, as well as the more prominent specimens of systematic moral theory, include principles of beneficence not limited to special relationships. Beyond a mere repetition of the principle that one ought to promote the good of others, what can be said to provide reasons for helping others to whom one has neither ties of affection, nor contractual obligations, to whom one has made no promise of aid, and whose need for help may not even be the result of one's own actions? In answer to this question, two main strategies have emerged. The first includes attempts to extend the model of intimate relationships to cover all of one's society or even all of mankind. The second includes attempts to develop grounds for general beneficence that do not rely upon any such extension of the model of intimate relationships but which assume, on the contrary, that ties of affection and sympathy are inevitably quite limited in scope.

Proponents of the first strategy sometimes say that we ought to behave beneficently toward all human beings because, as human beings, 'we are all members of the same family.' Since we are concerned here only with non-theological philosophical foundations for beneficence, I will not consider those forms of this argument which rely on the assumption that we are all one family because we are all children of God, and that God wishes his children to treat one another well. Instead, I will concentrate only on the secular version of what I shall call the family argument.[3]

The argument can be understood in several quite different ways. First, it may rest on the claim that we do have a special relationship with other human beings, different from our relationship to non-human animals. But this relationship might be either social or biological. If it is our social relationships which are said to make it appropriate to extend the model of familial relationships in such a way as to support a principle of beneficence to human beings in general, then the argument is weak, since we may in fact have little or nothing in the way of social relations with most other human beings, especially those in distant lands.

On the other hand, it might be said that it is the fact that we are all members of one species, one biological family, which grounds a general principle of beneficence. It is difficult to see, however, why the mere fact of membership in the same species should be so important, unless this is a shorthand way of basing a general principle of beneficence on the claim that we ought to promote the good of all human beings because human beings possess certain morally crucial characteristics. Different moral theories have attempted to specify different characteristics and then to ground a general principle of beneficence, along with other moral principles upon them, as we shall see shortly. But if this is the strategy, then it appears to be an abandonment of the family argument, rather than a development of it. For if the characteristics in question are universal to our species, they must be independent of the character of our actual special relationships to one another if and, this is the case, then it appears that the connection with familial relationships has been severed. Granted that this is so, it would be less misleading simply to dispense with the claim that all human beings are, or are like, members of the same family, and to proceed instead to examine the various moral theories which attempt to provide foundations for generalized beneficence which are not social, familial, or relational in nature.

Before doing so, however, it is important to note that there is a quite different way of employing the model of the family or of friendship or of loving relationships to provide foundations for generalized beneficence. Some strands of the socialist tradition have held that a fundamental change in the social order would make possible relationships among all persons – or at least all persons in the same society – which would share many of the attractive features of intimate relationships. According to this view, even though it is, and has been, true that human empathy or commonality of interest provides a basis only for severely limited beneficence, this is not an unalterable fact of human nature. Rather it is an artifact of defective modes of social organization. In a properly organized society, social institutions would serve to bind all of us together in a genuine community rather than in mere self-interested associations. This sort of view has, of course, been frequently labelled 'utopian' in the pejorative sense, on the grounds that the all-embracing extension of sympathy and concern it envisions is not psychologically possible, at least in populous modern societies.[4]

In contrast to the approach just sketched, the more traditional secular moral theories in the West have begun with the assumption that sympathy, affection, and commonality of interests can provide the basis for only a very limited beneficence. Within the confines of this essay, it will be possible only to sketch the more influential theories that rely on this assumption and to examine some of their implications for beneficence. We shall consider in turn: Kant's theory, John Rawls's ideal contract theory of justice as fairness, utilitarianism, Robert Nozick's version of a rights-based libertarian theory, and rational egoism as a theory of conduct which challenges all of the preceding views.

Though perhaps all systematic moral theories employ some sort of *universalization* principle at some level, Immanuel Kant's theory gives pride of place to this notion. According to Kant, the supreme principle of morality is the universal law formula of the categorical imperative, which directs one to act only on maxims one can at the same time will to be universal laws. A maxim, for Kant, is a principle which sets an end to be pursued or an action to be done, and which can or does serve to guide an agent's conduct. To will the universalization of one's maxim is to be willing for everyone to act on that maxim. In a famous example in *The Foundations of the Metaphysics of Morals* Kant uses the universal law formula of the categorical imperative to provide a justification for a principle of general beneficence.[5] He argues for a duty of general beneficence indirectly, by applying the universal law formula to a maxim of non-beneficence. His idea is that we are to ask ourselves whether we could will it to be a universal law that one may refrain from beneficence whenever acting beneficently would not maximize one's own interest. Kant concludes that one could not rationally or consistently will this, since to do so would be to will a world in which no one – including oneself – could count on disinterested aid from others. Though Kant himself does not make this clear, the success of the argument depends upon two very general but empirical premises: first, that the successful pursuit of one's ends, whatever they may be, depends on aid from others (and 'aid' here may include psychological support and love as well as the provision of material goods); and second, that one values the successful pursuit of one's ends more than one values total self-reliance or independence. If both of these assumptions are granted, it seems that Kant's universalization principle provides a plausible foundation for a principle of general beneficence, which he subsequently formulates rather cautiously as the imperative that one ought to help others in need, even when it is not in one's best interest to do so, at least so far as this can be done without excessive cost to oneself.

Kant recognizes several further limitations upon the practical implications of this principle. He emphasizes that beneficence is an imperfect duty, meaning that what the principle requires of us cannot be strictly specified: the principle itself tells us neither how much aid we are to render, nor what forms it is to take, nor even precisely toward whom we should direct our beneficence, granted that the finitude of our time and resources requires some selection. Kant attempts to ground the universal law formula itself in the concept of autonomy, which he in turn identifies with practical rationality. Kant's thesis is that an autonomous person is one whose will is determined by the idea of a law valid for all rational beings as such, and this idea is equivalent, he thinks, to the requirement of universalizability set forth in the categorical imperative. So the general duty of beneficence is ultimately based, according to Kant, not on concern for the well-being of others, but on respect for autonomy.

Kant distinguishes between duties of justice and duties of virtue, and contends that general beneficence, because it is an imperfect duty, must be classified among the latter. His view is that duties of justice command only that we observe certain 'outward' or behavioral requirements and that the prescribed behavior must be sufficiently specifiable that the duties may be enforced if necessary. Duties of virtue, in contrast, require that we set certain ends for ourselves, where setting an end is not reducible to the performance of certain specifiable acts which may be exacted from us. The implication, then, is that the duty of general beneficence involves something more than simply an obligation to perform certain acts (and Kant says that we are to make others' ends our own so far as this is possible); but also, in a sense, something less, since no definite list of beneficent actions can be spelled out.

It is not just that Kant's list of duties of justice does not include a duty of general beneficence: it also omits any more specific enforceable duty for promoting the good of others of the sort which establishes what we would now call welfare rights. Beneficence toward the poor or disabled, in Kant's theory, is a requirement of virtue, whose fulfillment is left up to the discretion of the individual or to voluntary associations of individuals who support charitable institutions. Kant believes that his moral theory implies what is now called a minimal or libertarian state – a government whose tasks are limited to enforcing valid contracts, and to protecting its citizens from physical injury, fraud, and theft. Positive promotion of our good in the form of welfare programs is beyond the legitimate scope of state authority and constitutes a violation of the rights of those who are compelled to contribute to their provision.[6]

In *A Theory of Justice* [12], and in subsequent published essays, John Rawls presents a normative political theory which he rightly describes as Kantian, but which arrives at conclusions concerning the relationship between beneficence and justice which conflict sharply with Kant's own.[7] Rawls uses the idea of a hypothetical social contract to generate the following principles of justice for regulating what he calls the 'basic structure' of society. The basic structure is the entire set of major political, legal, economic, and social institutions. In our society, Rawls notes, the basic structure includes private ownership of the means of production, the nuclear family, and the Constitution.

(1) The principle of greatest equal liberty: each person is to have an equal

right to the most extensive system of basic liberties compatible with a similar system of liberty for all (where the basic liberties include freedom of speech, freedom from arbitrary arrest, the right to hold personal property, and freedom of political participation, i.e., the right to vote, to run for office, etc.).

(2) The principle of fair equality of opportunity: persons with similar abilities and skills are to have equal access to offices and positions.

(3) The difference principle: social and economic institutions are to be arranged so as to benefit maximally the worst off.

According to Rawls, (1) is *lexically prior* to (2) and (2) is *lexically prior* to (3): we are first to satisfy all the requirements of (1) before going on to satisfy those of (2), and then those of (2) before proceeding to (3). Lexical priority permits no trade-offs when principles conflict; the lexically prior principle takes absolute precedence.

Rawls holds that the correct principles of justice are those that would be chosen by persons concerned to pursue effectively their conceptions of the good, and to do so under conditions allowing for the critical formulation and revision of those conceptions, in a situation of choice which satisfies certain conditions of procedural fairness.

Among the most important features of the hypothetical choice situation, which Rawls calls the 'original position' since correct principles of justice originate from it, is an informational constraint. The parties to the hypothetical contract are to choose from behind a 'veil of ignorance' so that information about their own characteristics or social positions will not bias the choice of principles. They are described as not knowing their own race, gender, socio-economic or political status, or even the particular content of their conception of the good. The informational constraints also help insure that the principles chosen will not place avoidable restrictions on the individual's freedom to choose and revise his conception of the good by insuring that the principles chosen will not be tied to any particular conception of the good.

Though Rawls offers several arguments to show that his principles would be chosen in the original position, the most striking is the *maximin argument*. According to this argument the rational strategy in the original position is to choose that set of principles whose implementation will maximize the minimum share of *primary goods* which one can receive as a member of society, and principles (1), (2), and (3) will insure the greatest minimum share. Primary goods are defined as those which are generally useful for the critical formulation, revision, and effective implementation of our conceptions of the good, and are said to include the basic liberties referred to in principle (1), the opportunities referred to in (2), and the various forms of wealth, broadly defined, covered by (3). Rawls's claim is that because these principles protect one's basic liberties and opportunities and insure an adequate minimum of wealth (even if one should turn out to be among the worst off), the rational thing to do in the original position is to choose them, rather than to gamble with one's life-prospects by opting for alternative principles.

Rawls makes it clear that the principles of justice are only a part of a larger *theory of the right*, which will include principles of virtue, one of which will be the principle of general beneficence. Though Rawls himself has not attempted to do so, he contends that the correct principles of virtue are those which would be chosen from the original position. In other words, the device of the hypothetical contract is supposed to supply rational foundations for both principles of justice and virtue, and presumably to provide a reasoned way of distinguishing between what justice requires in the way of promoting the good of others and what charity, or the virtue of beneficence, requires. Following Kant's way of making the distinction, the parties might reason that some principles governing their interactions should be enforced where necessary (principles of justice), while others (principles of virtue) are unsuitable for enforcement.

It should be clear that Rawls's theory does assign a prominent role to the promotion of the good of others *as a matter of justice* – in particular, the difference principle requires that we are to maximize the worst-off's prospects of certain primary goods. So quite apart from the requirements of a Rawlsian principle of the virtue of beneficence or charity, it looks as if beneficence plays a very large role in Rawls's conception of justice.

Rawls might protest, however, that it is misleading to view the difference principle as requiring that we promote the *good* of the worst off. For he does emphasize that his theory is *not* a teleological theory: it does not first offer a substantive conception of the good and then define the right (or the just) as that which maximizes the good. Rawls claims that the description of the original position assumes no substantive conception of the good: it is only said that the parties desire primary goods because they have a highest-order interest in being critical, autonomous choosers of ends. For this reason, Rawls emphasizes that he is not assuming that a person's share of primary goods is a measure of his *welfare* or satisfaction.

This reply, however, seems inadequate. For it can be argued that Rawls's characterization of the parties is based on one conception of the good rather than others. Though the idea that we have a highest-order interest in securing conditions that facilitate the critical formulation and revision of our conceptions of the good is more abstract than any particular conception of the good and though it does not identify a person's good with his welfare or satisfaction, it is a conception of the good nonetheless. Consequently, it seems quite appropriate to say that Rawls's theory assigns a very large role to beneficence at the institutional level, in that institutions are to be designed to promote maximally the good of the worst off, where their good is identified as their highest-order interest in securing conditions which allow them to flourish as critical, autonomous choosers of ends. And though, as we saw earlier, Rawls has not offered a contractual derivation of a principle of the virtue of beneficence, it is plausible to surmise that the content of this principle would also be determined, at least in part, by the ideal of the good on which the original position is founded.

Rawls makes certain controversial assumptions about the conditions for choosing principles which have striking implications for the scope of our obligation to promote the good of others. Perhaps the most important is his stipulation that the parties are to choose principles of justice on the assumption that these principles will be limited in application to the basic structure of the nation-state of which they are citizens. The implications of this restriction can be best appreciated if we contrast the consequences of implementing the difference principle within the confines of a prosperous, well-developed country such as the United States with the consequences of implementing that same principle on a global scale. Maximizing the prospects of the worst off in the United States certainly might require redistributive policies more stringent than those hitherto known here. But a theory which requires – as a matter not of charity but of justice, to be coercively enforced if necessary – that we are to maximize the prospects of the world's worst off, is perhaps the most revolutionary theory ever advanced. Rawls himself does not draw this radical conclusion; instead he merely sets aside the entire issue of distributive justice on a global scale, saying that the best strategy is to develop a theory at the national level, since it is within a nation that cooperative interactions are strongest.

Some of Rawls's critics, such as Robert Nozick, whose libertarian views we shall examine shortly, have charged that even when confined to a single country – even a very rich country – Rawls's difference principle confuses justice with charity. But if it should turn out that the reasons for adopting the difference principle are so general that it would be arbitrary to restrict the class of the worst-off to those living within one's own borders, then this charge becomes all the more forceful. The objection that Rawls has smuggled an ideal of charitable beneficence into the theory of justice may rest upon the intuition that even if justice requires that the prospects of the worst off be maximized until they reach some adequate level of the most important goods, it is certainly excessive to say that justice requires their prospects to be maximized *ad infinitum*, and irrespective of the costs or restrictions which this imposes on the better off.

There is another controversial stipulation in Rawls's description of the original position which again has far-reaching implications for the question of what justice, as opposed to the virtue of charity, requires in the way of promoting the good of others. Perhaps because he is aware of the exorbitant demands which certain health care needs may place upon social resources, Rawls stipulates that the parties are to choose principles of justice on the assumption that their needs fall within the 'normal range.' The suggestion seems to be that the satisfaction of extremely costly and relatively rare health care needs is not a matter of justice but of charity. However, as in the case of restricting the class of the worst off to one's own countrymen, Rawls offers no principled way of distinguishing special from normal needs.

Finally, the hypothetical contract methodology itself imposes a severe limit on the scope of principles requiring us to promote the good of others, whether they be principles of justice or of charity. Because the contract methodology views our moral relations with others as governed by principles which we could rationally agree on, it appears to be incapable of generating principles governing our treatment of nonrational animals. In particular, it seems unable to account for, in any obvious way, the widely held belief that morality requires that we should not wantonly inflict pain or death on nonrational animals, much less the belief that there are circumstances in which we should promote their good.

It is utilitarianism which Rawls views as his chief rival. Utilitarianism purports to be a comprehensive moral theory, of which a utilitarian theory of justice and of the virtues, including charity, are elements. There are two main types of utilitarian theory: Act and Rule Utilitarianism. Act Utilitarianism defines rightness with respect to particular acts: an act is right if and only if it maximizes utility. Rule Utilitarianism defines rightness with respect to rules of action and makes the rightness of particular acts depend upon the rules under which those acts fall. A rule is right if and only if general compliance with that rule (or with a set of rules of which it is a member) maximizes utility, and a particular action is right if and only if it falls under such a rule. On both versions, 'utility' is defined as pleasure, satisfaction, happiness, or as the realization of preferences, as the latter are revealed through individuals' choices.

It might seem that Utilitarianism would provide the strongest, most direct foundation for principles of beneficence, both in our special relationships and in our dealings with mankind in general. However, utilitarian thinkers have disagreed dramatically over the role which beneficence plays in the maximization of utility through individual conduct and at the institutional level.

Utilitarians usually draw a distinction between justice and virtues such as charity, holding that principles of justice are those whose utility is so great that they may be enforced if necessary. But utilitarians diverge sharply as to where the boundary should be drawn and as to the scope of beneficence both within and without the sphere of justice.

On the one hand, there are those, like the *libertarian* utilitarian Friedrich Hayek, who argue that the scope of beneficence as a matter of justice should be extremely limited [7]. Though it should provide its citizens with security of property and protection against fraud and physical harm, the state should not undertake positive efforts to promote people's welfare, except perhaps to establish a welfare 'safety net,' a minimal level of income or services below which no one should fall. There are two main types of utilitarian arguments for the claim that governmental beneficence should be kept to a minimum. First, it is argued that when the government undertakes to promote the good of its citizens, rather than simply to provide a secure framework within which they may pursue their own good as they see fit, various disutilities, including restrictions on individual liberty, inevitably result, both through abuses of power and well-intentioned miscalculation. Second, in its most effective presentations, this first argument from fallibility is supplemented with an economic analysis which purports to show that the operation of the free market is a more effective means of maximizing satisfaction or happiness. According to this analysis, the virtue of the free market is that it achieves the goal of beneficence through the most efficient allocation of resources, with the minimum of disutility resulting from government interference in individuals' choices, and it does so without requiring us to be benevolent.

Welfare state utilitarians, on the other hand, reject this line of argument on two grounds. First, they contend that even if a perfect market would achieve a more efficient allocation of resources than government efforts to promote our good, the markets with which we have to deal in the real world are afflicted with various imperfections. Moreover, they note that there are certain *externalities* which require governmental intervention to promote the good of the public or some portion thereof. An externality is present wherever the operation of the market produces costs or disutilities which cannot, within current arrangements, be taken into account by buyers or sellers operating in the market. The health-endangering effects of toxins discharged into a river by a chemical plant, for example, are not costs either to the producer, as producer, or the buyer, as buyer.

It might appear that governmental action to cope with externalities is more properly a case of preventing harm than of promoting good. However, in many such cases the force of such a distinction will be lost, granted a reasonable redescription of the activity as, for example, 'promoting health' rather than 'preventing illness'. But in any case, there is another class of phenomena, closely related to externalities of the sort just considered, which provides utilitarian grounds for positive governmental efforts to promote good, not just to prevent harm: so-called public goods. A public good is some benefit production of which requires the contributions of many but not all of the persons who will benefit from it, where the individual's contribution to it is a cost to that individual, and where it is either impossible or too costly to exclude non-contributors from the benefit if it is produced. In any situation of this sort there is the prospect that the 'free-rider' problem will block the cooperative effort required to produce the good. Each individual will reason that his own contribution or lack of it will have a negligible effect on the outcome – either enough others will contribute or they will not, regardless of what he does. But since his contribution is a cost and since he will be able to enjoy whether he contributes or not, the rational thing for him to do is not to contribute. If each person, or enough persons, reason in this way the good will not be produced.

The field of public health supplies numerous examples of this problem: large scale immunization campaigns and public sanitation projects are only two which come immediately to mind. Those who advocate an extensive role for government in promoting our good argue that compulsory participation in such programs may be necessary to overcome the free-rider problem.

Another argument advanced by those utilitarians who advocate a larger role for the state in promoting our good than libertarian utilitarians allow is based on the assumption that all persons have roughly the same capacity for enjoyment and that this capacity is limited. Thus, if we assume *diminishing marginal utility*, the principle of utility may direct us to move toward a more egalitarian distribution of wealth or goods or services, because the resources transferred to the worst off will yield more utility to them than the utility those same resources produce for the better off. Such redistribution for the sake of maximizing utility may take the form of various welfare programs designed to increase the poor's income or to provide greater access to health care, for example. Whether or not these benefits are to be considered rights, will depend, for the utilitarian, upon whether according them the special status that this implies will maximize utility. Rawls has argued that even if such a utilitarian argument for egalitarian measures is sound, it does not make utilitarianism plausible, because it commits the utilitarian to the intuitively implausible view that the commitment to equality is a purely contingent matter which depends entirely on the empirical assumption of diminishing marginal utility.

So far we have only considered the question of what sorts of *governmental* beneficence, if any, are justified from a utilitarian point of view, and what forms their justifications may take. And we have seen that different estimates of the effects of government policies versus the effects of self-interested interactions in the market lead to a utilitarian justification for the interventionist or welfare state, on the one hand, and for the minimal or libertarian state, on the other. The implications of Utilitarianism for the scope of beneficence at the level of individual conduct are no less controversial, since they too depend upon a wealth of empirical assumptions as to whether various actions will promote people's good and at what cost.

Though he departed from a libertarian utilitarian line in allowing some government intervention in the economy, in particular for the provision of certain public goods, John Stuart Mill argued that neither the state nor private individuals or groups should interfere in that part of a person's conduct that primarily affects only himself, even for the sake of promoting that person's own good, whether physical or moral. In his famous essay "On Liberty", Mill presents several arguments for this extremely stringent, anti-paternalistic limitation on private and public beneficence, and he claims to do so from an exclusively utilitarian perspective, explicitly eschewing any appeal to a notion of rights independent of considerations of utility. The strongest reason of all for adopting this anti-paternalistic principle, he says, is ". . . that, when [society] does interfere, the odds are that it interferes wrongly, and in the wrong place" ([9], p. 102).

From a utilitarian perspective, the force of this argument against paternalism – i.e., beneficence which involves interference with the liberty of the one whose good is to be promoted – depends upon the claim that there is great utility to be gained from individual liberty. There are three main ways in which individual liberty contributes to utility (and consequently three main reasons for not restricting it, even for the sake of promoting the individual's good) which can be gleaned from Mill's exposition. First, liberty itself – the freedom to choose and act according to one's choice – may be intrinsically satisfying. Second, liberty is of instrumental value in pursuing whatever other ends we happen to find intrinsically satisfying at a given time. And third, liberty – especially liberty of expression and association – is valuable as a condition for the critical formulation and revision of our ends over time, where the utilitarian's assumption is that we have greater opportunities for maximizing satisfaction if we are free to make informed judgments about what we will find most satisfying.

Several contemporary philosophers have argued that Mill's utilitarian case for an extremely strong anti-paternalistic restriction on beneficence is overstated. They contend that at most, utilitarian calculations show that there is a statistical presumption against imposing our beneficence on others, but that in any particular case or class of cases this presumption may be rebutted by empirical evidence about what will maximize utility. It is possible to interpret Mill, however, as holding that the evidence that interference in a person's primarily self-regarding behavior is disutilitarian is so strong that utility will be maximized if we operate within an absolute prohibition on such interference. The idea is that even if there are some exceptions to the empirical generalization that interference is disutilitarian, they are so infrequent, and so difficult to identify (except perhaps in retrospect), that utility will be better served if we proceed as if the strong statistical generalization were a universal truth.[8] It is still possible, of course, to reject Mill's position interpreted in this way and to offer utilitarian arguments for at least some forms of paternalism or imposed beneficence, by arguing that the generalization Mill advances is in fact *not* strong enough to warrant our treating it as a universal truth.

There is usually more agreement among utilitarians on the importance of non-paternalistic private beneficence. But even here there is room for debate about the most effective forms of beneficence based on differing estimates of the facts of the matter. The question of whether utility will be maximized through local or national charitable institutions, for example, must be answered by empirical research.

We have seen that because of their skepticism about the efficacy of governmental action and their confidence in the power of the market as an artifice for satisfying desires, some utilitarians opt for a libertarian or minimal conception of the role of the state. Others, however, argue for the libertarian state from an anti-utilitarian perspective. To avoid confusion we may distinguish between utilitarian libertarians, such as Friedrich Hayek, and *rights-based* libertarians, such as Robert Nozick [10]. As we saw earlier, rights for the utilitarian are not basic; they are derivative upon calculations of utility at the level of institutional design. For a libertarian such as Nozick, in contrast, rights-principles are the most basic moral principles, and their validity is wholly independent of considerations of utility.

This distinction is especially important for our purposes since the two types of libertarianism offer different foundations for principles of beneficence and for principles which place limitations on beneficence or upon the forms it may take. In particular, they provide radically different perspectives on the relationship between justice and charity, on the one hand, and between beneficence and autonomy, on the other.

Though there are many variants of rights-based libertarianism, I shall concentrate only on Nozick's, which may be the most prominent contemporary example. Although Nozick is primarily concerned to develop a rights-based libertarian theory of justice, in conscious opposition to both Rawls's position and Utilitarianism, what he says has important implications for beneficence. In Nozick's theory of justice, as in libertarian theories generally, the right to private property is a fundamental right, and it determines both the legitimate role of the state and the most basic principles of individual conduct.

Nozick contends that individuals have a property right in their persons and in whatever 'holdings' they come to have through actions which conform to (1) the principle of justice in initial acquisition and (2) the principle of justice in transfer. The first principle specifies the ways in which an individual may come to own hitherto unowned things without violating anyone else's rights. Here Nozick largely follows John Locke's famous account of how one makes natural objects one's own by 'mixing one's labor' with them or improving them through one's labor. Though Nozick does not actually formulate a principle of justice with respect to initial acquisition, he does argue that whatever the appropriate formulation is it must include a 'Lockean Proviso', which places a constraint on the holdings which one may acquire through one's labor. Nozick maintains that one may appropriate as much of an unowned item as one desires so long as (a) one's appropriation does not worsen the conditions of others in a special way, namely, by creating a situation in which others are no longer able to use freely (without exclusively appropriating) what they previously could or (b) one properly compensates those whose condition is worsened by one's appropriation in the way specified in (a). Nozick emphasizes that the Proviso only picks out one way in which one's appropriation may worsen the condition of others; it does not forbid appropriation or require compensation in cases in which one's appropriation of an unowned thing worsens another's condition merely by limiting his opportunities to appropriate (rather than merely use) that thing, i.e., to make it his property.

The second principle states that one may justly transfer one's legitimate holdings to another through sale, trade, gift, or bequest and that one is entitled to whatever one receives in any of these ways, so long as the person from whom one receives it was entitled to that which he transferred to you. The right to property which Nozick advances is the right to exclusive control over anything one can get through initial appropriation (subject to the Lockean Proviso) or through voluntary exchanges with others entitled to what they transfer. Nozick concludes that a distribution is just if and only if it arose from another just distribution by legitimate means. The principle of justice in initial acquisition specifies the legitimate 'first moves,' while the principle of justice in transfers specifies the legitimate ways of moving from one distribution to another. Whatever arises from a just situation by just steps is itself just.

Since not all existing holdings arose through the 'just steps' specified by the principles of justice in acquisition and transfer, there will be a need for a *principle of rectification* of past injustices. Though Nozick does not attempt to formulate such a principle he thinks that it might well require significant redistribution of holdings.

Apart from the case of rectifying past violations of the principles of acquisition and transfer, however, Nozick's theory is strikingly anti-redistributive. Nozick contends that attempts to force anyone to contribute any part of his legitimate holdings to the welfare of others is a violation of that person's property rights, whether it is undertaken by private individuals or the state. On this view, coercively backed taxation to raise funds for welfare programs of any kind is literally theft. Thus, a large proportion of the activities now engaged in by the government involves gross injustices.

After stating his theory of rights, Nozick tries to show that the state is legitimate so long as it limits its activities to the enforcement of these rights and eschews redistributive functions. To do this he employs an 'invisible hand explanation', which purports to show how the minimal state could arise as an unintended consequence of a series of voluntary transactions which violate no one's rights. The phrase 'invisible hand explanation' is chosen to stress that the process by which the minimal state could emerge fits Adam Smith's famous account of how individuals freely pursuing their own private ends in the market collectively produce benefits which are not the aim of anyone.

The process by which the minimal state could arise without violating anyone's rights is said to include four main steps. First, individuals in a 'state of nature' in which (libertarian) moral principles are generally respected would form a plurality of 'protective agencies' to enforce their libertarian rights, since individual efforts at enforcement would be inefficient and liable to abuse. Second, through competition for clients, a 'dominant protective agency' would eventually emerge in a given geographical area. Third, such an agency would eventually become a 'minimal state' by asserting a claim of monopoly over protective services in order to prevent less reliable efforts at enforcement which might endanger its clients: it would forbid 'independents' (those who refused to purchase its services) from seeking other forms of enforcement. Fourth, again assuming that correct moral principles are generally followed, those belonging to the dominant protective agency would compensate the 'independents,' presumably by providing them with free or partially subsidized protection services. With the exception of taxing its clients to provide compensation for the independents, the minimal state would act only to protect persons against physical injury, theft, fraud, and violations of contracts.

It is striking that Nozick does not attempt to provide any systematic *justification* for the Lockean rights principles he advocates. In this respect he departs radically from Rawls. Instead, Nozick assumes the correctness of the Lockean principles and then, on the basis of that assumption, argues that the minimal state and only the minimal state is compatible with the rights those principles specify.

He does, however, offer some arguments against the more-than-minimal state which purport to be independent of that particular theory of property rights which he assumes. These arguments may provide indirect support for his principles insofar as they are designed to make alternative principles, such as Rawls's, unattractive. Perhaps most important of these is an argument designed to show that any principle of justice which demands a certain distributive end state or pattern of holdings will require frequent and gross disruptions of individuals' holdings for the sake of maintaining that end state or pattern. Nozick supports this general conclusion by a vivid example. He asks us to suppose that there is some distribution of holdings D_1 which is required by some end-state or patterned theory of justice and that D_1 is achieved at time T. Now suppose that Wilt Chamberlain, the renowned basketball player, signs a contract stipulating that he is to receive twenty-five cents from the price of each ticket to the home games in which he performs, and suppose that he nets $ 250,000 from this arrangement. We now have a new distribution D_2. Is D_2 unjust? Notice that by hypothesis those who paid the price of admission were entitled to control over the resources they held in D_1 (as were Chamberlain and the team's owners). The new distribution arose through *voluntary exchanges of legitimate holdings*, so it is difficult to see how it could be unjust, even if it does diverge from D_1. From this and like examples, Nozick concludes that attempts to maintain any end-state or patterned distributive principle would require continuous interference in people's lives.

As in the cases of Utilitarianism and Rawls's theory, Nozick and libertarians generally do not limit morality to justice. Thus, Nozick and others emphasize that a libertarian theory of individual rights is to be supplemented by a libertarian theory of virtues which recognizes that not all moral principles are suitable objects of enforcement and that moral life includes more than the nonviolation of rights. Libertarians invoke the distinction between justice and charity to reply to those who complain that a Lockean theory of property rights legitimizes crushing poverty for millions. They stress that while justice demands that we not be *forced* to contribute to the well-being of others, charity requires that we help even those who have no *right* to our aid.

It is interesting to note that the incompleteness of Nozick's position as a general moral theory is more radical than its incompleteness as a theory of justice. In the latter case he offers rights-principles, draws the implication that the legitimate role of the state is quite limited and that neither individuals nor the state may interfere with a person's property even to promote his own good – but provides no foundations for these rights-principles themselves, though he does suggest that violation of them shows a lack of respect for the individual's autonomy. In the case of principles of morality other than rights-principles, including principles of charity or voluntary beneficence, he does not even formulate the principles. In fact his description of the hypothetical state of nature in which the correct moral principles are observed makes mention only of rights-principles.

Granted this lacuna, three problems loom large. First, even if Nozick is successful in developing foundations for his rights-principles, this will not be sufficient. These foundations must include or at least be compatible with a rational basis for principles of charity or voluntary beneficence. In particular Nozick must show that the fundamental respect for individual autonomy which he suggests grounds the theory of rights provides an adequate basis for, or is at least compatible with an adequate basis for, our shared belief that beneficence, where this means more than simply not interfering with others, is a central part of the moral life. Yet he must do so in such a way as to preserve his extremely sharp distinction between justice and beneficence. The difficulty lies in the fact that so far as Nozick suggests any basis at all for morality it is the idea of respect for autonomous persons, but his interpretation of respect seems to reduce it to the negative duty of noninterference.

It follows that Nozick must either introduce some other basic value into his moral theory – presumably concern for the welfare of others – or show that respect for autonomy requires a positive duty of beneficence while strictly forbidding enforcement of this duty. The difficulty with the first strategy is apparent: nothing in Nozick's theory as a whole or in his remarks about autonomy in particular provide any reason whatsoever why one should be concerned about the welfare of others. The second strategy is also problematic in the absence of a justification for Nozick's rights-principles, since it might be argued that if concern for the welfare of others is a basic moral value, it will place some restrictions on the scope of individual rights, including the right to private property.

Second, in suggesting that voluntary beneficence will ameliorate the suffering which would result from the system of private property in the absence of welfare programs, Nozick seems to assume without argument that individuals will be able to achieve their charitable goals by strictly voluntary association. This assumption ignores the fact that collective efforts for charitable ends may encounter public goods problems, if the success of the undertaking requires the efforts of a number of people. It is interesting to observe that the only examples of voluntary beneficence Nozick discusses are ones in which the benefit sought is not a public good, but rather one that can be provided by the efforts of a single individual. By overlooking the public goods problem, Nozick fails to address the charge that the humane results that he assumes will be achieved through voluntary beneficence may, in some cases, be attainable only through a larger role for government than he is willing to accept.

There is a third, more subtle difficulty to be faced by any view which, like Nozick's, attempts to soften the harsh consequences of a libertarian system of rights by assigning a significant role to voluntary beneficence. It is not enough to *say* that charity will flourish in a libertarian society, to the extent that, for example, those who cannot afford to purchase medical care in the free market will receive it through donations from the more fortunate. Nor is it enough to show that adherence to principles of charity is logically consistent with a libertarian theory of rights. It is also necessary to show that a social system governed by libertarian rights-principles – and in which the market plays a much greater role in interpersonal relations than at present – would nurture charitable attitudes, instead of breeding persons who are almost exclusively self-interested. Until each of these questions is answered, the libertarian's claim that his theory can accommodate the widely shared belief that charity is a basic element of the moral life will lack sure foundations.

Each of the moral theories considered thus far purports to provide reasons for acting or refraining from acting, even when doing so may not be in one's own best interest. Indeed, as we saw earlier, it is a common assumption of these competing views that a major task for moral theory is to provide principles which can ground duties to promote the good of others (or, as in the case of Nozick's theory, at least not to harm them), even where their good diverges from one's own. The doctrine of Rational Egoism, in contrast, not only purports to provide a different foundation for principles of conduct, including principles of beneficence, but to do so in a way that undermines all of the competing theories. Although formulating an adequate definition of Rational Egoism is not so simple as is sometimes assumed, a rough approximation must suffice for our purposes. As a normative doctrine, a prescription for how one ought to live (rather than as a psychological thesis describing the way people allegedly do, in fact, behave), Rational Egoism holds that everyone ought to act so as to maximize his own interest. The term 'rational' is an appropriate qualifier, since the claim is that it is rational to seek to maximize one's own interest.

Some discussions of egoism suggest a weaker formulation: pursuing one's best interest is always rational in the sense that one may always do so without being liable to rational criticism. This weaker version leaves open the possibility that one who does not pursue his best interest is not irrational, while making it clear that it can never be irrational to pursue one's best interest.

Two conclusions must be blocked if Rational Egoism is to be an interesting position. First, 'interest' must not be construed in a prejudicially narrow or crass way – for example, as limited to the interest in accumulating wealth or in partaking of the grosser sensual pleasures. Second, even though in one sense the content of 'interest' must be left open, this lack of restriction threatens to rob the notion of *self*-interest of all substance unless something is ruled out. The notion of self-interest must not be construed so broadly that any interest that a self may have is included. For then the distinction between egoism and altruism dissolves, since one's interests, on such an all-inclusive interpretation, may include an overriding concern to promote the good of others, even at one's own expense.

It seems that the following gloss of our earlier definition does much to avoid both the excessively narrow and the vacuously broad interpretations: Rational Egoism is the view that everyone ought to (or at least rationally may) maximize his own interest, regardless of how this affects the interests of others and regardless of whether his interest includes a direct concern for the good of others. For the most part I will for convenience operate with the simpler definition which states simply that it is rational for everyone to maximize his own interest, but the more complex gloss should be borne in mind since we shall be concerned with the implications for individual conduct and institutional arrangements which serve to promote the good of others.

We can begin with a rather crude version of the doctrine, which may be called 'Act Egoism'. It states that it is rational for everyone to maximize his own interest in a case by case fashion – the standard of self-interest is applied at the level of particular actions. On this view, my reason for acting or not acting beneficently, whether in special relationships or toward members of my own society or mankind in general, is always the same. Act egoism, then, does not proffer foundations for *principles* of beneficence, but only for particular acts of beneficence.

As in the case of arguments for Rule versus Act Utilitarianism, it has often been pointed out that the maximization of one's own good may best be achieved through adherence to rules which block appeals to the egoistic principle itself. Two basic types of argument are usually given for abandoning Act Egoism in favor of what may be called Rule Egoism. First, there are *fallibilist* arguments which emphasize that due to our bias, short-sightedness, and ignorance, there will often be situations in which a case by case evaluation of self-interest will not maximize one's interest. Second, there are arguments from *reciprocity* which emphasize that, in the long run at least, the best means of securing aid from others is to behave beneficently toward them. The second type of argument builds on the empirical psychological assumptions of the first. It is argued that the best long-term course is to act on principles of beneficence, rather than merely to try to appear to be beneficent, because, granted one's fallibility, the difficulties of maintaining a successful deception are simply too great.

This and related considerations might lead the egoist to change himself in more profound ways for egoistic reasons. An egoist might attempt to develop non-egoistic attitudes in himself not only in order to avoid the time-costs and errors of egoistic calculation, nor simply to secure the reciprocal beneficence of others, but also in order to become capable of new kinds of satisfactions that depend upon the existence of a direct regard for the good of others. For example, the egoist might conclude that persons who develop loving relationships founded on a direct concern for the good of others are capable of greater pleasure or fulfillment than those who do not.

Whether a perfectly rational egoist would regard principles of beneficence, either for special relationships or of a general sort, as mere rules of thumb to be abridged wherever a particular calculation of self-interest requires, or as virtually unchallengeable principles of conduct which he ought to internalize as fully as possible; and whether he ought to regard the promotion of the good of others as a mere means toward his own good or to strive to become the sort of person whose good is constituted in part by the good of others – all of these are purely contingent issues to be settled by an empirical investigation of the facts about the world in which the egoist finds himself. In this sense Rational Egoism as a foundation for beneficence is like Utilitarianism: the scope and limits of beneficence are not a matter of basic principles.

Rational Egoism, like Utilitarianism, may also be offered as a basis from which to derive principles of justice at the level of major institutions. Thomas Hobbes, for example, argued that what conventional morality regards as the most important principles of social justice are simply those rules of mutual restraint and mutual aid whose observance promotes everyone's most basic interest – the interest in survival ([8], Part I and II). Hobbes saw, however, that even if each recognizes that the general observance of principles of justice would be beneficial for all, it does not follow that these principles will be observed, because each individual, so far as he is a rational egoist, will attempt to be a free rider in the moral enterprise. Hobbes concluded that what is needed is a sovereign – a public power to enforce the rules so that compliance becomes rational for each of us. But Hobbes's critics have pointed out this solution to the free-rider problem cannot be achieved without creating an overwhelming coercive power that poses an even greater threat to the individual's interest than a general lack of compliance with moral principles.

The extent to which justice requires us to promote the good of others will depend, then, upon the contribution which enforced principles of beneficence make in promoting everyone's survival interests. Although he argued for virtually unlimited state authority to enforce whatever rules the sovereign deems necessary for making social life safe for rational egoists, Hobbes himself believed that a minimal, largely non-interventionist government policy, not one which undertakes massive welfare programs, would suffice. But again, as in the dispute between libertarian utilitarians and welfare state utilitarians, whether Rational Egoism provides a foundation for extensive or limited duties of beneficence, enforced or otherwise, is an empirical matter, not one to be decided by philosophical analysis.

Though I will not attempt to summarize this lengthy essay, a few concluding remarks are in order. We began with a recognition of the need for a theory of beneficence in health care, and soon the attempt to lay out the questions such a theory should answer led us to the conclusion that a more general theory of beneficence was required. The realization that beneficence must be understood in relation to other elements of the moral life widened the circle of our investigation even further. It became necessary to understand the place of beneficence within each of several competing general moral theories.

Though each of the theories considered can perhaps claim to give an account of the scope of beneficence and to ground a distinction between the virtue of beneficence and justice, there are striking differences. Both Utilitarianism and Rational Egoism (which is a normative theory even if not a moral theory) rely much more heavily than the other theories upon empirical data to determine the nature and extent of general and special duties of beneficence, the importance of benevolent attitudes, and the boundary between voluntary beneficence and enforceable beneficence. Though Kant's theory and Rawls's both purport to ground duties of beneficence in a more general moral theory based on the ideal of autonomous personhood, they arrive at opposing conclusions concerning the extent to which efforts to provide goods that are useful for autonomous persons may be enforced by the state.

In their present forms, both Rawls's theory of Justice as Fairness and Nozick's rights-based Libertarianism focus almost exclusively upon the question of whether, or to what extent, promoting the good of others is a requirement of justice. While neither theory now provides an account of the virtue of beneficence, the prospects for developing one seem much more problematic for Nozick, granted his tendency to equate morality with the non-violation of rights against physical interference or interference with private property.

Though none of these theories currently provides explicit answers to more specific questions about beneficence in health care, some differences do emerge in their implications for beneficence in the patient-provider relationship. Rational Egoism and Utilitarianism here, as in their responses to the question of whether the provision of health care may be achieved through coercion, assign a larger role than the other theories to factual premises. Rawls's theory and Nozick's, in contrast, imply that the proper role of beneficence in the patient-provider relationship depends ultimately upon non-derivative principles of justice that specify individuals' rights. For Rawls, the duties of the provider, like those of anyone occupying a social role, depend in the end upon the principles of justice that regulate that basic institutional structure within which roles are defined. For Nozick, a health care provider, like any person engaged in any occupation, may voluntarily choose to govern his behavior toward those he serves by principles of beneficence or even of self-sacrifice, and so long as both parties freely and knowingly consent to the arrangement, no one's rights are violated. But Nozick strenuously denies that the special character of the patient-provider relationship implies enforceable special duties of beneficence other than those that are generated in a particular case by an explicit contractual agreement.

To ascertain whether Kantianism, Rawls's theory of Justice as Fairness, Utilitarianism, rights-based Libertarianism, or Rational Egoism can provide an adequate account of the scope, limits, and foundations of beneficence has not been the aim of this essay. Instead, I have attempted to show that the search for answers to the most concrete practical questions concerning beneficence leads us into the domain of comprehensive moral theory, and that the adequacy of any proposed moral theory will in turn depend in part on its success in providing principled, consistent answers to these same concrete practical questions.

University of Minnesota–Minneapolis

NOTES

[1] See [4].
[2] See [11].
[3] Here I follow V. Haksar's usage in ([6], pp. 38–45).
[4] For a discussion of Marx's assumptions about the scope and limits of commonality of interest and of the family as a model of harmonious social cooperation, see ([5], especially Chapters I and IV).
[5] For a detailed reconstruction of Kant's argument, see [2].
[6] For an excellent introduction to Kant's ethical thought as a whole, see [1].
[7] My sketches of Rawls's theory, of Utilitarianism, and of Nozick's theory are adapted from a more detailed summary which includes references to the relevant texts in ([3], pp. 3–21).
[8] I would like to thank Rolf Sartorius for clarifying this point.

BIBLIOGRAPHY

1. Aune, B.: 1979, *Kant's Theory of Morals*, Princeton University Press, Princeton.
2. Buchanan, A.: 1977, 'Categorical Imperatives and Moral Principles', *Philosophical Studies* **31**, 249–260.
3. Buchanan, A.: 1981, 'Justice: A Philosophical Review', in E. Shelp (ed.), *Justice and Health Care*, D. Reidel Publ. Co., Dordrecht, Holland; Boston, U.S.A., pp. 3–21.
4. Buchanan, A.: (forthcoming), 'The Limits of Proxy Decision-Making', in R. E. Sartorius (ed.), *Paternalism*, University of Minnesota Press, Minneapolis.

5. Buchanan, A.: (forthcoming), *Marx and Justice*, Rowman and Littlefield, Totowa, N.J.
6. Haksar, V.: 1929, *Equality, Liberty and Perfectionism*, Oxford University Press, Oxford.
7. Hayek, F.: 1960, *The Constitution of Liberty*, The University of Chicago Press, Chicago.
8. Hobbes, T.: 1959, *Leviathan*, H. W. Schneider (ed.), Bobbs-Merrill Company, Inc., Indianapolis.
9. Mill, J. S.: 1956, *On Liberty*, C. V. Shields (ed.), Bobbs-Merrill Company, Inc., Indianapolis.
10. Nozick, R.: 1974, *Anarchy, State and Utopia*, Basic Books, New York.
11. Rawls, J.: 1955, 'Two Concepts of Rules', *The Philosophical Review* 64, 3–32.
12. Rawls, J.: 1971, *A Theory of Justice*, Harvard University Press, Cambridge, MA.

WILLIAM K. FRANKENA

BENEFICENCE IN AN ETHICS OF VIRTUE

Health care is usually discussed in recent medical ethics under the heading of rights and/or justice, as it is in the companion volume to this one. It can be and is also treated under the heading of duty or obligation. In these approaches typical questions are, respectively, (1) Is there a right to health care?, (2) Is the present distribution of health care just?, and (3) Has society, or some segment of it, a duty to provide health care? It is clear that these three approaches may overlap or even coincide, e.g., justice is sometimes subsumed under duty or under respect for rights. In this volume, however, health care is being dealt with under the rubric of beneficence. The idea, I take it, is that doing so involves a different perspective from those of rights, justice, or duty. Actually, this is not necessarily the case; beneficence may, for example, be thought of as a right on the part of the beneficiary or at least as a duty on that of the benefactor. In Cicero, Seneca, and Kant it is conceived of as a duty, though without any belief in the existence of a corresponding right. A different perspective is entailed only if one takes an ethics of virtue approach to beneficence and sees health care as falling, at least in part, under beneficence. These points are illustrated to some extent by other essays appearing here. My own assignment is to look at beneficence as it would appear in an ethics of virtue. Such an ethics has often been attributed to Plato and Aristotle and is sometimes advocated in recent thinking about morality.

This means that my essay must be relatively abstract and theoretical. But, although most of it has a wider application, it is written with questions about health care (HC) in mind, in the hope of being philosophically illuminating to those concerned about such questions. I do not, however, write as an advocate of a virtue ethics conception of morality; I have actually discussed such a conception of morality several times, on the whole negatively, though I am now less opposed to it than I was then.[1] Whether this qualifies me as an authority on beneficence in an ethics of virtue (EV) remains to be seen. In any case, I am not here concerned to argue for or against an EV approach to beneficence (and HC) as much as to show what it is like.

I

Behind the conception of this essay lie two contrasts. In a familiar one,

Earl E. Shelp (ed.), Beneficence and Health Care, 63–81.

beneficence is contrasted with justice, whether they are thought of as virtues or as duties (or as rights), and HC is regarded as falling under either of them or possibly under both. In a more general, perhaps less familiar, one, an EV is contrasted with an ethics of duty (ED). It is more general, because an ethics of each kind may include both beneficence and justice as virtues or duties respectively. I shall bring in the first contrast on occasion, but must at once say something about the second, especially about the nature of an EV as compared with an ED.

An ED or an EV may appear in two forms: as a normative ethical theory espoused by a moral philosopher or as the morality of a culture or society. In either case, in an ethics or morality of duty the concept of Duty or of the Right, and the correlative concept of the Wrong, are taken as basic or primary, and those of Virtue and Vice, or Moral Goodness and Moral Badness, as derivative or secondary. More accurately, what is basic or primary in an ED is judging certain external actions or kinds of actions in certain circumstances, e.g., telling the truth, treating people equally, or doing what will promote the greatest general good, as right, duty, or obligation, and the opposite as wrong, independently of the motive or disposition involved, and then judging the disposition to perform such actions in those circumstances as *therefore* morally good or bad, a virtue or a vice. The actions are right, etc., because of what they are, do, or bring about in the world, independently of the state of mind or trait involved, and the state of mind or trait is morally good or virtuous only *if* and *because* it involves trying to do what is *right,* etc. Virtues and vices have, or at least may have, an important place in an ED, but what is basic is rightness of action or duty.

The contrasting nature of an EV is roughly indicated by a recent comic strip in which one character tells the other, "You should stop being such a bully. It isn't a very attractive personality trait." This comic also illustrates how natural it is for us to use EV manners of speech. In an ethics or morality of virtue, matters are approximately the reverse of what they are in one of duty. What is basic or primary is the concept of certain dispositions, motives, or traits as 'attractive,' morally good, or virtuous (or the opposite), independently of the external actions they issue in. These actions are then judged to be good, virtuous, or right (or the opposite) *because* they are such as issue (or would or would not issue) from those dispositions, motives, or traits, not because of what they are as such, or because of their results. What one *is* or means to *be* is what matters, not what one *does*; or rather, what one does matters, not as such, but only because of the disposition or kind of person involved. A disposition or trait is not a virtue because it includes a concern to do one's duty or what is right in the way of action, as most EDs would have it (indeed, it may be a virtue, even though it does not include such a concern); rather, an action is right or what one ought to do only if and because the disposition involved is a virtue.

This last claim is crucial to an EV. That the *goodness* or *virtue* of an action depends on its motivation is standard doctrine, even in an ED like Kant's. The question is whether the *rightness* or *wrongness* of it depends on its motivation. An EV may claim that an action is right if and only if it is actually motivated in a certain way, regardless of what it is, does, or brings about. But this claim is not easy to defend, and EVs often take the more plausible line of holding that actions are right or wrong if and only if they are what a person who was motivated in a certain way, e.g., by benevolence, *would* do or try to do if he or she knew what they were about.

In an ED, then, the basic concept is that of a certain kind of external act (or *doing*) to be done in certain circumstances; and that of a certain disposition's being a virtue is a dependent one. In an EV the basic concept is that of a disposition or way of *being* – something one has or is, not does – as a virtue, or as morally good; and that of an action's being virtuous or good, or even right, is a dependent one. Other kinds of ethics are also possible. In one, the basic concept is that of a *Right* or *Rights* one person has in relation to others; in it the others are thought of as having duties if and *because* the one has a right or those rights, and dispositions are regarded as good or as virtues if and *because* they involve a respect for such rights. Thomas Reid called such an ethics of rights (ER) a system of natural jurisprudence because he perceived it as created in the image of a legal one. Such an ER would be the reverse of an ED in the sense that, if an ED gives an important place to rights, as it may and in modern times usually does, it will regard A as having a right in relation to B if and because B first has a duty toward A (and later I shall tend to mention rights as well as duties in connection with EDs), but an ER, if it gives a place to duties, as it may and usually does, will regard B as having a duty toward A if and because A first has a right in relation to B. In a fourth kind of ethics, the basic concept would be that of *the Good*, conceived as something to be taken as an end to be brought about or promoted; in it actions and dispositions would be held to be right, good, or virtuous if and because they bring about this good or involve trying to. Ethics of these four sorts may overlap or even coincide in practice, but they also may not. Our own moral ethos, as Sidgwick pointed out, contains at least the germs of all of them.

II

More about the nature of an EV will come out if we see what an EV says or may say about beneficence or benevolence. But we must observe that an EV need have no place for such a virtue; Plato in the *Republic* did not, and even Aristotle recognized it only in so far as it was involved in liberality (the getting and giving of things whose value is measured by money) or in friendship (which for him is not really altruistic).[2] Except perhaps for the Stoics, beneficence as now understood was hardly an important virtue in the West before the advent of Christianity. Even if an EV recognizes it as a virtue, however, it may not regard it as a cardinal or basic one; it certainly need not regard benevolence as the only basic one or as the whole of virtue. The view that benevolence is "a supreme and architectonic virtue, comprehending and summing up all the others, and fitted to regulate them and determine their proper limits and mutual relations" has been prominent in moral philosophy in modern times as a result of the influence of the Christian ethics of love,[3] but Hume, who was propounding an EV, did not hold it, regarding benevolence as just one among others, including justice, even though he maintained that our moral judgments are all, directly or indirectly, an expression of a sentiment of humanity or sympathy. In what follows, I shall assume that an adequate EV must posit at least one cardinal virtue besides beneficence/benevolence, namely justice. The question is: What does or should such an EV say about beneficence/benevolence?

What is beneficence/benevolence? It is not just a certain kind of action, like walking, telling the truth, or distributing things equally. It is a habit, tendency, or trait of *character* that one has (rather than of 'personality,' as the comic has it). It is a disposition, a way of being rather than of acting; it does manifest itself in action (and perhaps in feeling), but not in any one definite kind of action. It is obviously related to doing good, not doing harm, preventing or remedying evil, etc., but these things can be attempted or done in many different ways. Beneficence/benevolence is something that has a goal but may use various means of achieving it. Furthermore, 'it' is at least two things; two different, if related, dispositions or traits must be distinguished, namely, beneficence and benevolence. Benevolence is not just well-*wishing* or well-*meaning*-ness, as Kant pointed out; it is that, but it is more, and includes well-*willing* and genuinely *trying* to do good to others. Being beneficent entails actually doing or producing good and not evil, and one may by inadvertence, ignorance, or sheer bad luck, succeed only in bringing about evil even when one is genuinely seeking to do otherwise.

Bertrand Russell once wrote an essay entitled "The Harm That Good Men Do," ([8], pp. 109–120) and had no trouble finding examples. Benevolence does not necessarily entail wisdom, but even if one argues that true benevolence will try to get the wisdom it needs, one cannot equate it with beneficence. One who is benevolent and knows what to do may still lack the ability or power to bring about the good he or she wills.

Again, something can be benefi*cent* without being bene*volent*, for example, the sun or the climate. Even a person or an institution can have a disposition that is beneficent without being benevolent. Out of self-interest, for example, a man might cultivate tendencies to act in ways that in fact promote the happiness of those about him. He would then be beneficent, 'a good guy to have around,' but he would not be benevolent. For him beneficence would merely be the best policy, as a proverb says honesty is. A beneficent trait is not necessarily a benevolent one or vice versa. The 'doctor' in G. Donizetti's *L'Elisir d'Amore* claims to benefit humanity and appears to do so, but his motives are self-serving.

There are then two dispositions or traits that might be regarded as virtues but must be distinguished. They are, however, connected. One can say that benevolence is a disposition to be beneficent and that beneficence is what a benevolent person would be or have if he or she had the needed knowledge and power. Even so one can be or have one without the other. Then just what disposition is it that an EV should consider to be a virtue, or more exactly, a moral virtue? Clearly not beneficence. For being beneficent requires having ability or power of a sort that a morally virtuous person may in fact not have and need not have in order to be virtuous; and, besides, as we saw, being beneficent does not entail being benevolent. Beneficence is neither a necessary nor a sufficient condition of being virtuous even in the respect here in question. However, being benevolent is certainly a necessary condition of being virtuous in this respect; actual benevolent motivation is at least part of virtue for any EV that believes in beneficence/benevolence at all. Is it a sufficient condition of having the virtue we are trying to identify? Hardly. To have this virtue in its fullness one must not merely be benevolent, one must also have, so far as is in one's power, the other things needed for doing what one wants, *qua* benevolent, to do. This involves at least having something like what Aristotle called practical wisdom, a habit of thinking clearly and logically and of knowing or trying one's best to know the facts that are relevant to one's decisions – particular facts, scientific generalizations, maybe even metaphysical or theological truths. Borrowing a term some theological moralists have used in a somewhat different way, one might say that the virtue we are trying to pinpoint is a combination of *benevolence* and *responsibility.*[4] Our title word 'beneficence' is hardly the word for this virtue, but neither is 'benevolence.' I shall therefore call it BR. It is then BR that I think an EV does or should regard as a virtue.

Some who agree that the virtue in question should not be called beneficence will contend that it may still be called benevolence, arguing that one is not *really* benevolent unless one also has what I called responsibility. This, however, strikes me as pushing; 'really' is a *really* tricky word in such contexts. I shall therefore use 'BR' for what I am talking about, but I shall not mind much if others call it beneficence or benevolence, so long as it is clear that they are talking about BR and not mere beneficence or mere benevolence. I must, however, observe that BR is not to be thought of as a combination of mere benevolence and mere beneficence. It is, no doubt, true that, ideally, the benevolent man would be beneficent as well as benevolent, but this involves his having the cooperation of fortune or providence, and so goes beyond the requirements for being morally virtuous.

BR is not just a tendency to act; it includes thinking in the ways just indicated, and presumably having certain feelings and emotions in relation to others. Its central component, however, will be a certain kind of motivation, namely, to do good and not evil to others. Now, one might be moved to avoid doing harm to others or to do them good because one thinks this to be right or to be one's duty. Then one's BR would be part of one's disposition to do one's duty or what is right, at least as one sees it. This is how BR is conceived in an ED, e.g., by Kant. Kant distinguished between 'pathological' and 'practical' benevolence. In the latter one is moved, not by a direct concern for the good of others, but by a sense of duty; in the former one is moved by such a direct concern for others, one feels love for them or simply desires not to harm them, to help them, etc., independently of any thought about this being right or obligatory or of its being a right they have. And Kant insisted that benevolence as a moral virtue can only be 'practical' benevolence; 'pathological' benevolence may be very well, and there is nothing abnormal about it, but it cannot be a duty and it is not a moral virtue. But an EV cannot take this view of its BR, for doing so entails making duty basic to virtue. It may even hold that BR so conceived is not a good thing, not a virtue, as the poet Schiller implied when he wrote that on Kant's view one must despise others entirely, and then with aversion do what one's duty enjoins one. Today proponents of EV are likely to contend that a morality emphasizing such a 'conscientious' kind of BR would generate 'alienation,' which many follow Marx in regarding as the worst thing anything can do.

If BR in an EV cannot be 'practical' in Kant's sense, then what is it? Well, it might be the sort of benevolence Kant calls 'pathological', plus what I called responsibility. Indeed, this is precisely how J. D. Wallace conceives of "primary benevolence," which he represents as a cardinal virtue and pictures in EV terms. For him 'primary' BR is not a form of "conscientiousness", i.e., of a concern to act rightly; it is or involves "a *direct* concern for the happiness and well-being ... [or] *good* of others," i.e., a concern for their good that is not mediated by any belief about its being a duty or even a virtue. It is or involves a straight-out taking of the good of others as an end, independently of making any ethical or value judgment about doing so.[5]

Now, like Kant and Prichard, I am myself doubtful that BR, thus conceived, is a *moral* virtue, though it is, of course, a good thing; but most EVs would insist that it is, as Wallace does. In any case, one may think of such direct BR in different ways. One may hold that direct benevolence is a disposition we naturally have, at least in germ, as Hume did, even if one thinks it is not a moral virtue unless it is cultivated and deliberately developed into BR by us, as Aristotle did. Or one may hold, as Jonathan Edwards did, that we do not have it naturally, since we lost it in the Fall; and that it is a disposition which now comes only through a rebirth due to God's grace – and only to some of us. One may also take some intermediate view, e.g., that there is a natural 'element of the dove' in us, or at least some of us, but that true charity or *agape* is supernatural or infused, though still capable of being developed by cultivation; or that even direct benevolence is an acquired disposition, but one acquired by a wholly natural process of conditioning or reinforcement and, once acquired, capable of being cultivated as a trait of character and a virtue. All of these views of BR are open to proponents of EVs.

There is, however, another possibility – that there is a form of BR which is not 'direct' or 'pathological' and yet is different from Kant's 'practical' benevolence. Wallace writes,

> Someone who is deficient in [direct concern for the good of others] ... might admire [benevolent] people for their [benevolence] and want as far as he can to be like them. He might then want to do in certain situations what a [benevolent] person would do. Acting as a [benevolent] person would act *because one regards* [*benevolence*] *as a virtue* and wants, *therefore*, to emulate the [benevolent] person is meritorious, and it reflects credit upon the agent. ... [It] involves taking as one's ideal the (primarily) benevolent ... individual, and undertaking to act as he would act. The predominant motive in so acting is the desire to be good oneself – to be as much like a good person as possible. This is different from the motive in primary benevolence, which is a [direct] concern for the good of another. I do not mean to suggest the agent has an *ulterior* motive in actions

> . . . characteristic of [such] secondary benevolence – one might simply take the benevolent person as *worthy of emulation because of his goodness* . . . and resolve to be more benevolent [than one is] ([11], pp. 133, 156f).[6]

Wallace calls such benevolence 'secondary' because "It depends, for its merit, upon the fact that primary [benevolence] *is* a virtue and is thus a worthy ideal at which to aim" ([11], p. 133). Actually, I think such a secondary BR might take a purely 'pathological' form; one might simply want to be like the primarily benevolent person without first thinking of him or her as being good or virtuous, i.e., without making any ethical or value judgment about being like that, and hence without being motivated by a desire to be *good.*[7] But, as the italicized words show, Wallace is conceiving of his secondary BR as being mediated by a belief that primary BR is (morally) good or virtuous. It is not a form of conscientiousness, since it is not mediated by a belief that acting benevolently or beneficently is a *duty*, but it is like conscientiousness in being indirect or non-'pathological', i.e., in presupposing some kind of ethical or value judgment.

There are then at least two kinds of BR that may be urged as virtues by an EV: direct BR and the form of BR Wallace regards as secondary. Most EVs have stressed the former and ignored the latter, as Hume, Foot, and Wallace do.[8] But I am not sure that their neglect of the latter is justified. They neglect it because they view it as presupposing the judgment that direct BR is good or a virtue. But even if it does, one might still hold that direct BR is not *morally* good, not a *moral* virtue. A woman who is directly BR is benevolent simply because she wants to promote the good of others, not because she believes promoting their good is good, right, or virtuous. As directly BR she does not act under the aegis of any value judgment, let alone a moral one, and it is therefore not clear that she can be morally good or have a moral virtue. One may then plausibly choose to espouse an EV that emphasizes the latter kind of BR, i.e., a disposition to be like the directly BR person except for being moved at least in part by what Hume called a regard for virtue instead of being moved only by a direct concern for the good of others. Even if the second sort of BR presupposes that the first sort of BR is good or a virtue, it does not presuppose that it is a moral virtue or morally good. There are dispositions or traits that are good but not morally good, or virtues but not moral ones. It certainly does not presuppose or entail that a directly BR person is better morally than an indirectly BR person. At any rate, one can reasonably argue that one's virtue is not *fully moral* unless one's primary BR (or whatever) is associated with a secondary BR, i.e., is judged by oneself to be a virtue and pursued as such.

III

An EV, then, cannot conceive of a BR as a form of conscientiousness, as an ED or ER would; it usually regards BR as being or involving a direct or 'pathological' concern for the good of others, but it may emphasize the possibility of an indirect or, in Kant's sense, 'practical' concern for their good, and perhaps it may place the two kinds of BR in some kind of tandem arrangement. But, whatever an EV does in this respect, we may still ask what such an EV says *about* BR as it conceives it. It must make some kind of moral judgment about its BR, but it cannot say that BR is a duty, or a right, for to say that is to fall back into an ED or ER. It must make, as a basic premise, the judgment that BR is morally 'attractive', morally good – a moral virtue. It can say that we ought to be BR, but only as an inference from this premise, not the other way around. But, in holding that BR is a moral virtue, it is not necessarily asserting that BR is beyond the call of duty or supererogatory, for an EV cannot recognize any call of duty as basic or ultimate, but only a pull or attraction of goodness or virtue. An EV can also hold that we ought to be benef*icent*, but only because it is what a BR person would be if he or she could, not because it is a duty in some independent sense.

We must also notice that EVs can give various justifications or rationales for their basic judgment or principle that BR (or some other disposition) is a moral virtue, e.g., that BR is good or a virtue because God prizes it, because it makes for the greatest general happiness, or because it is needed for human flourishing. An EV may also claim that we simply intuit as self-evident that BR is good or a virtue, or that we see it as good or virtuous when we take the moral point of view or are in a certain original position or in reflective equilibrium.

Perhaps I should put in a word here about benevolence and justice. One could conceive of benevolence as a general sort of goodwill that would include being just to others, keeping promises, telling the truth, etc. Taken in this broad sense benevolence could well be held to include the whole of virtue.[9] I am assuming, however, that benevolence is a concern, direct or indirect, for the good of others, i.e., not to harm them, to do them good, prevent evil from coming to them, or removing, remedying, or compensating for the evil that comes to them. Then benevolence views conduct under the aspect of the good or evil actually or intended to be done, prevented, etc., and not, except incidentally, under those of promises kept, truth told, etc. Justice, on the other hand, sees conduct mainly under the aspect of how matters such as goods and evils, promise keepings, truth tellings, etc., are distributed, e.g., whether people are treated equally, or whether they are being treated as they deserve or according to their rights. It is not concerned about conduct or character simply on the score of producing, preventing, or remedying good or evil, benefit or harm, but rather on the score of the comparative treatment of the persons (or other centers of experience) involved.

IV

Let us suppose, then, that we have an ethics or morality in which BR is put forward as a virtue for each of us to assume if we do not have it and to exercise if we have it – as a cardinal virtue along with others such as justice. The picture this suggests is that each of us is to cultivate BR and then in each situation do what one's BR moves one to do. As Augustine put it, "Love, and do as you please." But things are not quite that simple, because we are supposing that there are other cardinal virtues, and it may be that sometimes implementing one is incompatible with implementing another. Mercy may season justice on one occasion but justice may temper mercy on another. The benevolent or rather BR act may not always be what the good or virtuous person would do. I shall largely ignore this complication but it must not be forgotten. Even if BR calls for a deed of health care in a certain case, that deed may still not be the one to do.

But, if we assume that BR would include a concern for the health of others, as seems plausible (the 'doctor' in the opera referred to earlier supports his claim to benefit humanity by alleging, among other things, that he restores his customer's health), then our picture suggests a society of individuals each of whom has a concern for the health of the others and does what he or she can about it, at least in so far as this is consistent with other benevolent concerns, with being just, etc. There would be something ideal about this, but it does not seem as if this can be the whole answer to the question about beneficence, or rather BR, and HC. The problem about BR and HC in contemporary medical ethics is not simply one about what being BR would lead each of us to do for others in the way of HC. Individuals are not, as such, in a position to do much about the health of others, except in certain limited situations, and it seems clear that the problem of HC would not be adequately resolved just by our having BR as individuals.

So far we have been talking as if our question were simply one of personal ethics, i.e., of the ethics governing the relations of persons to persons as such (what Kant called the metaphysical principles of the doctrine of virtue), considered independently of the special relationships they may also be in.

Actually, in our society there are individuals who, because of some role, office, vocation, or special circumstance in which they are, have or should have a special concern for the health of others, either in general or in some particular cases, and it looks as if this is how it should be even in a society pervaded by an EV in which BR is plugged as a cardinal virtue. We must therefore ask about the operation of BR in the contexts thus set up. There seem to be five sorts of these contexts:

(1) those of parents in relation to their children in youth, of children in relation to parents in age, of teachers and educational administrators in relation to their pupils or students,

(2) those of friends and lovers in relation to one another,

(3) those of employers in relation to employees,

(4) those of medical doctors, nurses, and hospital administrators in relation to their patients and the public, and

(5) those of people in government and in social agencies of various sorts concerned, directly or indirectly, with health in relation to the public.

A recent article ([1], p. 3) raises the question whether cooks and restauranteurs ought to undersalt their food out of a concern for the health of their patients, refuse to serve high cholesterol foods, etc. (apparently it was written by a gourmet maverick on the staff, for it answers that health is not the highest value in such situations!); perhaps, then, we should add another kind of context:

(6) those of cooks and restauranteurs, food manufacturers, etc., in relation to their clients and customers, at least if an ounce of prevention is worth what the proverb says it is.

All of these contexts are such as to offer a place for HC, but there are important differences. In most, the positions involved are chosen voluntarily, at least in societies like ours, but in some they are not, e.g., that of being a child, a pupil in the lower grades, or a member of the public (I suppose being a patient is a matter of choice, much as one has a choice when a robber says "Your money or your life"). In some, the positions, at least on one side, are directly concerned with HC, e.g., those in (4) and some of those in (5); in the others they are more or less incidentally concerned with HC, as in (3) or (6).

Now, in a society so set up as to provide these contexts, one naturally thinks of some people involved as having duties or obligations and others as having rights, hence a moral philosopher concerned to advocate an EV might be opposed to society being organized thus, taking a morality of purely personal virtue, including BR as earlier described, to be the whole story. That is, he might condemn the kind of 'coherence' John Donne lamented the loss of in the lines

'Tis all in pieces, all coherence gone; All just supply, and all relation: Prince, subject, father, son, are things forgot . . .

As I intimated, however, it is hard to believe that HC – and other such concerns – could be adequately provided for in a society relying only on purely personal virtue, and we must therefore ask whether an EV must take the line suggested and how it may conceive BR as functioning in the contexts listed. In most of the contexts in question it is hard to eliminate talk about the duties and the rights of those involved, but perhaps this can be accommodated within an EV, as we shall see in a moment.

Given then a society with such contexts – such 'coherence' and 'relation' – as ours has, what parts may BR play in it from the point of view of an EV? In the first place, it seems clear that an EV could regard those contexts, or rather the offices, positions, and roles involved, as providing channels for the operation of whatever personal BR individuals may have. Unchanneled BR would be at least partly blind, and result in much scattering of effort. But the existence of the family, for instance, provides a place within which it may take shape, somewhat as CARE and UNICEF do. Indeed, part of the ideal of the family seems to be that, if each couple provides for its children, children will in general be benefitted, more than they would be if each couple tried directly to help provide for them all. In fact, some people might deliberately choose one of the positions mentioned, e.g., those in context (4), precisely in order to have a more defined way of exercising BR. It could even be argued that society should create such channels for the exercise of BR and other virtues. The net effect, of course, would be that a given individual's BR toward some persons might be diminished as that toward yet others is increased.[10] But there would still be plenty of room for a more undirected BR, at least in any incompletely structured society such as ours.

Secondly, the proponent of an EV might well argue that some of the contexts involve positions naturally calling for a special cultivation of BR such as is not called for in others, especially (1), (4), and (5). Socrates even held in opposition to Thrasymachus that a ruler *qua* ruler will be concerned for the good of his subjects, but it is more plausible to hold the corresponding belief about parents, doctors, and nurses. It goes without saying about friends. Having just read Donne's *Songs and Sonnets*, I fear that lovers are a mixed bag, but our EV would no doubt urge BR on them too, not as a duty but as a virtue.

(4) is a special context for BR. Here the individuals involved on the agent side (as distinct from their patients) have voluntarily chosen a role or vocation of which the object is to provide HC. This does not mean that they have BR or choose their vocation out of BR; they may choose it because of the money, the status, and the kind of life made possible. Indeed, it has been contended that they have no duty to provide HC and their patients have no right to it – that one can and should think of the HC enterprise as just another private industry, except that the service or product it sells is HC, and as being subject to no laws, principles, rules, or values other than those regulating other such businesses. This point would hold, if at all, only for private HC agencies and their employees, and does not touch the question whether public HC agencies ought to be established. In any case, an EV is, as such, concerned to represent BR (and the HC entailed) as a virtue and not as a duty or right, hence it can in a sense sidestep this issue. That BR is a virtue for special cultivation in HC situations is not easy to deny, or, for that matter, in context (5) cases as well.

V

Thus far my discussion of BR in an EV has been a consideration of it as a virtue of individuals, simply as such or as occupants of certain special positions. In connection with context (5), however, one remembers Plato's view that there are two analogous sets of virtues, one for or in the individual and another for or in the state or society as a whole, the latter not being reducible to the former though perhaps consequent to them. Benevolence was for him not one of these virtues, but here we must consider the possibility of BR as a virtue for or in a society or state, where this is not simply reducible to the BRs of its members, or even of its governing officers, either in private life or in social office. A United States senator from Michigan recently said that public policy must be humane and cooperative in relation to the auto industry. If we take him to mean that public policy should be BR, then we have the idea of BR as a kind of public virtue. Perhaps we should also have corporations in mind. A corporation president was reported lately as saying that corporations should learn and practice justice, truth, fairness, and democracy; he did not include benevolence in his list, perhaps purposely, but let us add it, thinking especially of corporations related to HC.

Now, it is hard to see how an institution like a corporation or state can be literally benevolent or have BR; it is not a center of motivation and thought in the required way; only persons are, and corporations and states are hardly persons except for certain legal purposes and then only as fiction. Benevolence and BR proper can reside only in the individuals who hold corporate or governmental offices. Clearly, however, a corporation or government can be *beneficent* in the way distinguished earlier, and so we may, and perhaps must, conceive of its 'virtue', not as benevolence or BR, but as *beneficence*, i.e., an actual tendency not to do harm, to do good, etc. In an EV this would be thought of as a tendency to do or bring about what the BR person would do or bring about, if he or she could do what a corporation or state can do. A corporation or state might have this tendency through its policies even if its officers were not motivated by BR but it would seem more likely to have it if they were; either way, it would be bene*ficent* and hence at least quasi-virtuous.

For a state or nation to be beneficent, its laws, its other institutions, and its policies, including those relating to HC, must be beneficent in their operation, either directly or through their effects on the character and conduct of individuals. One might think, as did the moral philosopher of a few pages ago, that an EV must content itself with touting such things as BR as morally good or as virtues and cannot itself generate laws and institutions regulating conduct, that it can consistently, at most, merely recognize the existence of such laws and institutions and seek to foster and express BR and other virtues as best it can, given their existence. It is important for us, at this juncture, to see that this is not so. The point is that a philosopher or a society committed to an EV takes as basic the goodness of certain dispositions like BR (i.e., takes them as ideals), and, seeing that there is not enough of these dispositions in enough people, may seek to ameliorate the situation by enlisting certain kinds of aids. Among these would be laws requiring, prohibiting, or simply permitting certain sorts of action in such contexts as (1), (3), (4), (5), and (6). As Bishop Robinson puts it, the virtues "may, and should, be hedged about by the laws and conventions of society, for these are the dykes of love in a wayward and loveless world," e.g., laws such as those being proposed or opposed lately in connection with HC and medicine ([7], p. 115). Robinson mentions 'conventions' as well as laws, for it is not just laws and legal institutions that may be advocated and set up by a philosopher or society espousing an EV. There is also what has been called *positive social morality*, i.e., the adoption of a moral code to which non-legal or moral sanctions like praise, blame, approval, disapproval, esteem, ostracism, etc., are attached. Such a code may overlap with the legal code of a society, but it may be for or against kinds of conduct that are not properly in the province of law. The sanctions may be either external, as when one person judges, praises, or blames another, or internal, as when one judges, praises, or blames oneself through the operations of conscience, feels guilt and remorse, etc.

Such a positive social morality (PSM) might itself take an EV form, i.e., the form of saying and thinking things like "That wouldn't be benevolent," "He is unkind," "Arrogance is a vice," etc. And, with respect to BR, it may take at least three forms. The first would involve expecting and/or praising BR and its exercise in action and blaming its absence, at least in situations like HC in which BR is relevant. In the second, BR and its exercise would be praised, but its absence would not be blamed or even expected. The third would be more complex: there would be certain areas in which BR would be expected and/or praised and its absence blamed, and other areas in which it would be praised but not expected or its absence blamed. The ancient Greek PSM is often held to have had an EV form of one of these sorts. Plato has Protagoras describe it as follows:

> ... if a man is wanting in ... good qualities ... and has only the contrary evil qualities, other men are angry with him, and punish and reprove him. Of these evil qualities one is injustice, another impiety ... the child cannot say or do anything without [parents, nurses, and tutors] telling him that this is just and that is unjust; this is noble, that is base; this is pious, that impious; do this and don't do that. If he willingly obeys, well and good. If not, he is straightened by threats and blows ... later they send him to teachers, and enjoin them to see to his manners ... and the teachers do as they are asked ... When [children become adults] the state again compels them to learn the laws and live after the pattern which they furnish, and not after their own fancies ... He who transgresses them ... is called to account ... (*Protagoras*, 323d–325e).

Here Protagoras somewhat blurs the distinction between a PSM and the law, as the Greeks were wont to do, but he does seem to see the Greek PSM as an EV, and my quotation makes it look as if he sees it as a PSM of the first sort just described, though the omitted portions of his speech also suggest one of the third sort. In any case, my quotation shows that a virtue PSM may be very much like a duty one in practice. It is of interest to observe that Protagoras does not list benevolence as a good quality or its opposite as an evil one.

ED forms of PSM can also be of various sorts, of course, but in them duty-talk and perhaps rights-talk, too, will be basic, not virtue-talk such as Protagoras describes. Our own PSM, if we have one, seems to be a mixture of these sorts of talk, as a glance at our discussions in medical ethics shows. Or perhaps one should say that our culture is pluralistic and involves more than one kind of PSM both as to form and as to content. However this may be, I have been saying that a philosopher or society with an EV ideal of morality may advocate or generate a corresponding form of PSM, indeed it seems natural for him or it to do so. But I want also to point out that he or it may come out with a PSM involving judgments of duties and rights, made *as if* they were basic, much as they are in law, which clearly subordinates aretaic or virtue-talk to deontic talk or talk of duties, rights, justice, etc. The reasons for this would be much the same as those that are behind having a legal system. In such a PSM health care might be represented as a duty and/or as a right, but the basic premises on which the PSM rests would in this case not be premises about what is duty or what is a right, but about what is virtuous or good in a way of character or disposition. It would only be thought that, for moral education and regulation, a PSM is necessary in addition to law, and that a deontic PSM is more effective than an aretaic one. For such a thinker or society, the ideal might include that of being a person who is BR and does *what* the BR person would do *as* he or she would do it, independently of the rules of any PSM, and possibly departing from those rules on occasion, as Socrates, Jesus, and Martin Luther King did. It would, however, be thought that, human nature and the world being what they are, this ideal must in practice be supplemented, not only by a system of law and legal sanctions, but by a PSM of rules about duties and rights accompanied by moral sanctions, for those too young, too impulsive, or too un-benevolent to live by the ideal. And it might well be thought that such a PSM, as well as the law, should say something about HC. In fact, the PSM (and the law) might even assign special duties and rights of beneficence and HC in type (4) contexts, just as it would to parents or teachers. In short, from the bed of an EV ideal there may grow a PSM that is just like a PSM rooted in an ED or ER in vocabulary, sanctions, etc., and *within* the PSM it may even be that duty and rights seem basic to virtue, even though the PSM as a whole, as viewed from the *outside*, is nourished by a basic EV ideology. This point has not been sufficiently noticed by friends or foes of EVs; in fact, the friends are almost always vehemently against the 'rules' and 'legalism' of ED forms of PSM, and sometimes seem to be against having any kind of PSM whatsoever.[11]

It turns out, then, that there is an important way in which an EV view of basic morality can accommodate the talk about duties, obligations, and rights that is so natural in most of the contexts listed earlier, in connection with HC as well as with other matters. It need not forget 'prince, subject, father, son' – or doctor, patient.

VI

There is much more that one could say about BR in an EV in relation to HC, but I shall take up only one more topic, related to the foregoing one, namely that of 'professional ethics'. A professional ethics is a kind of mini-PSM adopted by or prevailing among the members of a profession, in this case the members of the HC profession or of some branch of that profession, and possibly accompanied by special sanctions. It need not be and usually is not just an encoding of principles or values holding for moral agents as such, whatever their vocation. It will normally overlap with the law and the more general PSM of the inclusive society, but it may rightly contain articles differing from or even contrary to them, e.g., a clause of greater confidentiality. The ethics of a HC profession may also put a special store on BR in its members, beyond that expected of others. For example, one of the precepts associated with the Hippocratic oath, sometimes taken as a paradigm of the professional ethics of the physician, admonishes physicians "to help, or at least do no harm."[12] A professional ethics may take either an ED or an EV (or an ER) form, and it may take an ED even where the underlying ethics is an EV, just as a more general PSM may. If it does, it might make HC a duty or a right, even if the law and the PSM of the larger society do not, e.g., it might specify that a doctor has a duty and his patient a right, in a way in which others do not, because of their roles in relation to one another. Notice that the design of these roles allows a doctor to treat people unequally in this sense, as well as to restrict or channel his beneficence at least to some extent. Here again the idea seems to be that, with each doctor having this special BR toward his patients, everyone gets roughly equally well treated in matters of HC, at least if they can afford it or are subsidized; and back of this idea, in turn, there seems to be the idea that everyone has a right to be someone's patient, i.e., that the society and the profession should so dispose things that an adequate doctor and hospital is available to everyone, with or without subsidy. It seems to me that this is how the situation should be, whether one's basic ethics is an EV or some other, though my last remark may not be consistent with a pure EV. In any case, the articles of a professional HC ethics cannot be determined satisfactorily merely by considering the interests of the profession involved (or of its patients); they must also in general reflect and be consistent with the law and the PSM of the larger society, at least where these are themselves justified; though, as was indicated, they may depart from them on some points; and, in particular, like that law and that PSM, they must ultimately be judged by the basic principles of morality, whether these are principles of duty and rights or principles of virtue.

Looking back, I have identified three levels at which an EV may operate in discussions of beneficence/benevolence and health care: (1) that of basic ethics or morality, where the EV with its virtue of BR may serve as a guide to personal being and doing; (2) that of a PSM which takes an EV form, as it may; and (3) that of a professional ethics which takes an EV form, as it too may. As we saw, however, it is also possible, on an EV foundation, to erect a PSM or a professional ethics that has an ED or ER form. The basic EV will then serve as the ultimate ground for determining whether there should be a PSM or professional ethics or not, what form it is to take, and what its content should be.

The University of Michigan
Ann Arbor, Michigan

NOTES

1 See [5], pp. 184–160; [3], pp. 21–36 and [2], pp. 63–70.
2 Plato's ethics was an EV, but I am not sure Aristotle's was.
3 See [9], p. 238.
4 Here responsibility or being responsible does not presuppose having duties or include a sense of duty or respect for rights.
5 See [11], pp. 128ff, 133, 157. Because he counts conscientiousness as a virtue, I would regard him as combining a partial ED with an EV. But see also David Hume in *An Enquiry into the Principles of Morals* (many editions), and [4], pp. 305–316.
6 The italics are mine, and I have put in "[benevolence]" where he uses "generosity". He regards generosity as one form of benevolence.
7 One's desire might then *generate* judgments of goodness or badness, rightness or wrongness, *à la* Stevenson, but it would not be *predicated upon* any such judgment.
8 Actually, Wallace in a sense recognizes *three* forms of BR; one is a part of conscientiousness, the others are primary BR and secondary BR. Foot in the paper cited seems to be arguing for an EV of direct BR and other such direct concerns, cf. [4].
9 See [9], pp. 242f.
10 Or, some sorts of beneficence would be supererogatory for one person, but not for another, depending on their roles.
11 So-called new moralists have argued variously (among other things):
(a) for an EV versus an ED,
(b) for not having any PSM,
(c) for having a PSM of EV form,
(d) for merely changing the rules or virtues of our PSM.
12 See [10], p. 20.

BIBLIOGRAPHY

1. Anon.: 1981, 'Ethical Issues in Haute Cuisine', *The Hastings Center Report* 11 (February), 3.
2. Frankena, W.: 1973, *Ethics*, 2nd ed., Prentice Hall, Inc., Englewood Cliffs, New Jersey.
3. Frankena, W.: 1973, "The Ethics of Love Conceived as an Ethics of Virtue', *Journal of Religious Ethics* 1, 21–36.
4. Foot, P.: 1972, 'Morality as a System of Hypothetical Imperatives', *Philosophical Review* 81, 305–316.
5. Goodpaster, K. E. (ed.): 1976, *Perspectives on Morality*, University of Notre Dame Press, Notre Dame, Indiana.
6. Kant, I.: 1959, *Foundations of the Metaphysics of Morals*, transl. by L. W. Beck, Liberal Arts Press, New York.
7. Robinson, J. A. T.: 1963, *Honest to God*, SCM Press, Ltd., London.
8. Russell, B.: 1928, 'The Harm That Good Men Do', *Sceptical Essays*, Allen and Unwin, Ltd., London.
9. Sidgwick, H.: 1901, *The Methods of Ethics*, 7th ed., MacMillan and Co., London.
10. Temkin, O., *et al.*: 1976, *Respect for Life*, Johns Hopkins University Press, Baltimore.
11. Wallace, J. D.: 1978, *Virtues and Vices*, Cornell University Press, Ithaca.

JOHN P. REEDER, JR.

BENEFICENCE, SUPEREROGATION, AND ROLE DUTY

I. INTRODUCTION

When people are sick, they need medical treatment; as we often say, they need help. But why do doctors and other medical personnel aid the sick? Is their work an expression of some general duty of beneficence? Is there a form of beneficence which is not a duty but supererogatory? What is the relation of beneficence to other duties?

I will propose that there is a general duty of beneficence (mutual aid) and that there is also a type of beneficence which is supererogatory; the duty of medical personnel, however, is role specific. I will try to sketch a theory of the relation of mutual aid, supererogatory beneficence, and specific role duties. A number of writers have insightfully charted parts of the terrain, but the map still needs work. We can shape this landscape, of course, as we map it: I report my judgments (and those of others) and suggest a normative account.

In the next section I will discuss mutual aid as part of a system of moral requirement. But why have a system of *requirement* at all? Why not rely on our natural concern for others, our altruistic benevolence? Some thinkers, of course, have suggested that we simply do desire the good of others as well as our own; we desire and act for the good of everyone (defined as a particular unit or all humankind). Our collective benevolence, if you will, is the foundation of morality. (Bene*volence*, as I use it, names a disposition to desire and act so as to increase good and decrease evil for human beings; bene*ficence* signifies a principle or ideal of conduct).[1]

My assumption, however, is that benevolence should not serve as our starting point. Our concern for others is part of our moral experience, but we also need a system of moral requirement whose motivation is independent of benevolence. Benevolence serves as an additional motivation for doing what is required, and, as I will suggest below, it can express itself in supererogation, but we should devise a system of duties and rights which is motivated independently of our good will to others.[2]

Such a system would in the first place establish general prohibitions against injury and coercion and it would provide principles of justice for the major

Earl E. Shelp (ed.), Beneficence and Health Care, 83–108.

institutions of social life: sex and family, economy, and political order. In particular, for the economic sphere, it would establish principles for the distribution of benefits and burdens involved in the appropriation of natural resources and the creation of a social product. The system would also provide a principle of mutual aid: in certain situations, those in a position to do so have a prima facie duty to render aid at some level of cost to themselves. We might speak of the principle as 'beneficence', but it would be a requirement. It would not spring from our natural concern for the good of others or the good of all.

II. MUTUAL AID

To define a principle of mutual aid, we need to specify (1) the sorts of situations in which it applies; (2) the sorts of aid to be rendered; (3) the cost to the giver; (4) the relation of mutual aid to other duties; and (5) procedures for selecting givers and receivers.

(1) The principle of mutual aid, as I define it, does not apply where there are appropriate institutional structures (based on the principles of justice). I assume that the system for the production and distribution of the social product would provide for human needs which the ordinary operation of the economy fails to meet, e.g., assistance to the unemployed. We believe, roughly speaking, that the economy ought to provide for those unable to work or unable to meet their needs, and we establish a principle of 'welfare' or 'relief'. Were this not our conviction, then mutual aid might extend (as it has in some cultures) to cases of economic deprivation. In addition, our system of political order would provide structures for preventing crime and aiding victims.

What then is the sphere of mutual aid? First, there are natural disasters such as famine, flood, fire or earthquake; there are the natural risks, say, of swimming or running. Now the general distribution of income and wealth, and hence the ability to acquire food, clothing, shelter, medical care, and other goods necessary to meet basic needs can seriously affect one's capacity to respond to natural disasters. But we do seem to distinguish between the poverty caused by social injustice and the poverty and consequent starvation which might accompany, for example, a severe drought in a normally fertile area.

In addition to natural risks or disasters, there are also accidents with cultural artifacts. People build machines, for example, and use them in various endeavors for the production and distribution of goods and services; we use machines even for play. But machines sometimes malfunction or the individuals in control of them either through negligence or error endanger others.

Finally, there are cases where individuals should help others get institutional prevention or amelioration or step in when institutional help is not forthcoming. The neighbors watching from windows who failed to call the police during the attack on Kitty Genovese did not help her get institutional protection or assistance. In addition, if institutional protection is not presently available, one might, e.g., depending on the cost one is willing to incur, have to defend a Kitty Genovese from her attackers. If no institutional aid to victims is available, then this task as well might fall to individuals. There is, for example, the case in the New Testament. When Jesus teaches that to "inherit eternal life" you must love God and "your neighbor as yourself", a lawyer or scribe asks a further question:

> But he, desiring to justify himself, said to Jesus, 'And who is my neighbor?' Jesus replied, 'A man was going down from Jerusalem to Jerico, and he fell among robbers, who stripped him and beat him, and departed, leaving him half dead. Now by chance a priest was going down that road; and when he saw him he passed by on the other side. So likewise a Levite, when he came to the place and saw him, passed by on the other side. But a Samaritan, as he journeyed, came to where he was. and when he saw him, he had compassion, and went to him and bound up his wounds, pouring on oil and wine; then he set him on his own beast and brought him to an inn, and took care of him. And the next day he took out two denarii and gave them to the innkeeper, saying, "Take care of him; and whatever more you spend, I will repay you when I come back." Which of these three, do you think, proved neighbor to the man who fell among the robbers?' He said, 'The one who showed mercy upon him'. And Jesus said to him, 'Go and do likewise' (Luke 10: 29–37).

According to some interpreters this parable teaches that the 'neighbor' you are to love is anyone in need; others see it proclaiming that you pay whatever is necessary; still others see the parable's emphasis on the Samaritan's compassion.[3] In fact, we might take the story to illustrate compassion and altruistic benevolence, not mutual aid. But even if the Samaritan acted out of altruistic benevolence the point of the story seems to be that he did what the principle of mutual aid required.[4]

Two important qualifications need to be added if we are to have an adequate notion of the sorts of situations to which mutual aid is applicable. First, an individual may have been careless or taken a special risk and thus exposed himself or herself to some natural, mechanical, or criminal danger, e.g., going swimming in an area known to be especially hazardous. Or perhaps an individual has received aid, e.g., in a famine, but squandered it.[5] In contrast, the paradigm of a mutual aid situation is one in which the victim is *not* at fault. We may, however, want to extend the principle to cases where the individual is at fault, with the proviso perhaps that aid will be rendered at a lesser level of cost; we might treat second offenses as outside the scope of the principle altogether.

Second, in order to have a situation to which the principle of mutual aid applies, the one in a position to give is not ordinarily thought to be at fault, to be morally responsible for the need arising. For example, it is one thing to cause a car accident where you have not been negligent about maintenance or driving, or even made an error in judgment; your brakes suddenly fail, despite the fact that you have them regularly checked. Here the principle of mutual aid seems to apply and it seems irrelevant that you yourself were part of the causal chain which precipitated the accident. Error or negligence, however, would seem to change your moral relation to the victim and call for some form of compensatory action which might be more costly than mutual aid would require.[6]

Mutual aid, then, is applicable to certain threatening or dangerous situations, naturally, mechanically, economically or criminally caused, where neither victim nor rescuer is ordinarily at fault.

(2) Second, what is at stake in these situations is life, bodily integrity, and the satisfaction of basic needs through food, clothing, shelter, and medical (including psychological) care; let us call these states welfare, in contrast to the satisfaction of other desires which we can call well-being.[7] The principle of mutual aid is not designed to further the well-being of individuals even if it is frustrated by the same occurrences that threaten them in more basic ways. This rough concept of welfare would furnish us with a guide to the sorts of help mutual aid provides.[8] In some situations you are to prevent harm or suffering of these sorts; in others you are to render positive aid if harm has already occurred.[9] For example, you should pull me out of the way of the volcano's lava, but if I have already been burned, then you should try to get my wounds attended to (in the most efficient way possible).

(3) Third, there is the question of cost. Some thinkers have proposed that mutual aid be rendered only at a minimal cost or risk to the giver; others suggest that we should be willing to aid up to the level of a comparable cost to ourselves, we should help up to the point at which we would suffer an equivalent sort of deprivation or harm. I would personally prefer the latter, but its justification is doubtful. Let me try to explain, however, what it would mean.[10]

If I am required to aid others in regard to their welfare, then it is plausible to take 'at no comparable cost' to mean that I should be prepared, if necessary, to sacrifice my well-being in part or in whole. I am not only required to change my life-style – my use of income and wealth according to my preferences and choices – but my choice of vocation, what Michael Slote means I believe by life-plan, something he asserts we are not required to sacrifice.[11] Even if it is true that we attribute a great deal of *value* to the exercise of the capacity to choose a vocation (or a style of life), this value is not as important (in my scheme) as welfare;[12] furthermore, mutual aid overrides my right to this choice just as it overrides my right to use (in morally acceptable ways) my justly acquired income or wealth.[13]

It would be one thing, of course, to postpone or alter one's vocational plans for a short time, but on the construction of 'at no comparable cost' I am working out, one is required to make alteration for as long as necessary: I should give up my plan to build wooden boats in a Rhode Island boatyard in order to work as a paramedic in an area of chronic famine. One can imagine a society where one would be legally required to do so, unwillingness being regarded as an intolerable selfishness. Furthermore, 'at no comparable cost' would also seem to require that I sacrifice my choice of personal relationships or even a particular relationship. For example, I might desire marriage but marriage is impractical for paramedics in famine stricken areas; I might want to marry Betty but she does not want to be married to a paramedic.

In addition, beyond these elements of my well-being, I am required to sacrifice my welfare up to the point of loss equivalent to what I give. For example, suppose a drowning swimmer whom I have tried to save grabs my shoulders; I can't extricate myself and a proper save is impossible. My life is not in danger, but to rescue the person I will have to put my feet on sharp coral in order to steady us both until additional assistance arrives; the cuts on my feet will mean that I will never walk or run normally again (if I were a professional athlete, my career would be ruined, but I am assuming we are supposed to sacrifice that). Quite apart from the fact that I have tried to rescue the individual and have perhaps preempted someone else from making the attempt, the interpretation of 'no comparable cost' I suggest would require that one sustain these injuries in order to save life.

Thus my suggested interpretation of no comparable cost not only ranks welfare above well-being, it assumes a relative ranking of life over limb. This conception of mutual aid would also rank bodily integrity in relation to the satisfaction of basic needs; it could require that I go hungry even to the point of sustained malnourishment, to protect you from severe injury. But although the ranking of life over limb might meet with general agreement, the valuation of basic needs in relation to noninjury is likely to produce disagreement. Is being uninjured really more important than being healthy? Any conception of mutual aid costs rests on value judgments and a justification of the principle will have to take into account that what is a 'minimal cost or risk' to one might be major to another.[14]

I will simply note the two final issues to be considered, but not suggest a solution:

(4) We would have to decide whether mutual aid overrides or is overridden by other duties. For example, do I have a special obligation to my children, say, such that I am justfied in meeting their basic needs in full while other children starve to death? Aiding an accident victim overrides my promise to take you sailing, but are there promissory duties which are nondefeasible?[15]

(5) There are also distributive issues about givers and recipients:

(a) Who should aid when more than one can? When some can do so more effectively and at less cost than others? When some fail to do anything and thus increase the level of what is needed from others?[16]

(b) What if more need my aid than I can satisfy? Should I aid some adequately and others not at all? How do I choose? Or should I distribute aid equally, albeit at an inadequate level? Should I withhold some resources now in order to help others later?

Now how might we justify a principle of mutual aid? The sorts of situations it applies to are extraneous to institutional provisions and thus require separate attention. Let us assume that the system of requirement based on the principles of justice provides for opportunities to achieve well-being; the system corrects the natural and social lottery in ways we deem just and it does not require that some sacrifice their justly acquired well-being to increase the well-being of others. Why then would we agree to a requirement of mutual aid and how would we decide on the level of cost? Bracketing the relation of mutual aid to other duties and distributive issues about givers and receivers, what we need to justify is mutual aid at a particular level of cost.[17]

Gilbert Harman suggests that our allegiance to a principle of mutual aid rests on an actual convention:

> Hume says that some, but not all, aspects of morality rest on 'convention.' There is a convention in Hume's sense when each of a number of people adheres to certain principles so that each of the others will also adhere to these principles. I adhere to the principles in my dealings with others because I benefit from their adherence to these principles in their dealings with me and because I think that they will stop adhering to these principles in their dealings with me unless I continue to adhere to the principles in my dealings with them ([15], p. 103).

On this view, mutual aid is an actual convention or agreement, reached, as Harman says, through a complex process of "implicit bargaining and mutual adjustment" ([15], pp. 104, 111). It is an exchange based on self-interest, a matter of reciprocal justice.[18] Harman suggests that the principle of noninjury is also an actual convention but we might hold that noninjury and the principles of justice for the distribution of the social product are based on some other foundation. Mutual aid could still be conceived as an actual convention, a fair exchange of my acceptance and adherence for yours.[19] I assume that we do not believe it is in our mutual interest to agree to a convention which would require us to promote the well-being of others; most of us believe, however, that it is to our advantage to agree to a principle of mutual aid.

Compliance to whatever principle is agreed on, however, will not be perfect. Accepting the principle commits one to trying to live up to it but we anticipate moral weakness and sheer inadvertence. We do not estimate the likelihood of these to be so great as to render the adoption of the principle futile. We might agree, however, that consistent moral failure (to help others) would disqualify one from receiving aid.[20] Not agreeing to the principle at all of course (or not agreeing to it at a certain level of cost) would place someone outside the convention altogether. As Slote puts it in his 'Brazen Rule',

> It is not wrong to omit doing something for others that others would have omitted doing for you, if your positions had been appropriately reversed ([28], p. 135).[21]

But those who do adopt the convention are agreeing to yield their rights to income, wealth, or even perhaps bodily parts (however these rights are justified) in order to secure compliance to mutual aid. For example, consider this case discussed by Dan Brock (in an argument against a utilitarian interpretation of aiding others) [5]. Would we be willing, asks Brock, to yield our right to life and let ourselves be killed so that our organs (the only suitable ones) could be used to save two other lives? We would agree, says Brock, only if we knew that we were one of these three individuals but we didn't know which; if we knew that it was in fact our organs that were needed, we would not consent. Now since we do not know that eventualities will put us in the first situation, we will not agree to a principle that could cost us our lives. As one of three, my chances of needing an organ over against losing my life are two to one; but what are the probabilities in the general population? In this situation of uncertainty, I might be willing to yield my right to a bodily part up to the point of a comparable cost because I value life above having a full complement of organs and I am willing to risk losing, e.g., one kidney, to get others to agree to give me one if I need it. I am willing to agree to give one of my two good kidneys so that someone else can have one good one but I am not willing to die so that one individual can have my heart and another my kidney. What I want is the best protection I can get, while I exist; I would not bargain with my life unless it is clearly in my interest to do so (I *am* one of the three, but don't know which one).

Now assuming the general importance of welfare and the unpredictable nature of disasters and accidents, everyone has reason to want *some* principle of mutual aid. But those who are well-off may not feel they need to agree to a principle which requires aiding others at great cost to themselves, as 'no comparable cost' does. Those who accept the convention are reasoning from an actual situation which will not be in perfect compliance with the principles of justice. While it would be possible to reason from perfect compliance, my suggestion is that we adopt mutual aid in light of what really obtains. As injustices are rectified or increased, the meaning of mutual aid could change. They might be willing to agree to yield their rights to a certain amount of wealth or income, but they would not agree to a principle which required the sacrifice of vocational choice or a bodily part. They will be able to protect themselves in any situation so it is in their interest to agree only to a principle which requires minimal cost or risk to the giver. Harman indeed sees the inequality of wealth and power as the reason we have a strong principle of noninjury and a weak principle of mutual aid:

For, whereas everyone would benefit equally from a conventional practice of trying not to harm each other, some people would benefit considerably more than others from a convention to help those who needed help. The rich and powerful do not need much help and are often in the best position to give it; so, if a strong principle of mutual aid were adopted, they would gain little and lose a great deal, because they would end up doing most of the helping and receive little in return. On the other hand, the poor and weak might refuse to agree to a principle of noninterference or noninjury unless they also reached some agreement on mutual aid. We would therefore expect a compromise .... ([15], p. 111).

Now it is not necessary to agree with Harman that noninjury is an actual convention, the product of a bargaining process, to see the plausibility of his reasoning in regard to mutual aid itself. The well-off have reason to want a principle but not a costly one; they can hire lifeguards.[22]

But even for the well-off matters are not so clear. In the first place, wealth itself is precarious. Secondly, in the situations in which one needs aid one's provisions may not work, e.g., one's lifeguard may be ineffective. In regard to some goods, e.g., kidneys for transplants, one may have to depend on mutual aid or the benevolence of others. Actual ignorance of the future – the precariousness of wealth and the sheer unpredictability of circumstances – might approach the ignorance of the original position in a hypothetical contract. Nonetheless, the well-off can protect themselves better than the poor and they will estimate this when deciding on a principle of mutual aid.

Consider the fact, however, that even the rich have to estimate. To agree to a convention of this sort, one makes an empirical judgment about the likelihood of future events and states and wagers accordingly. Different individuals even with the same class interests might disagree on which principle – which level of cost – is to their advantage. One makes a judgment about what one is likely to need and about what others are likely to agree to give.

Moreover, there might well be differences in the risks that individuals are prepared to take. A strong principle of mutual aid – up to comparable cost – is an insurance policy for which we pay a high premium. That we need the aid and that others are in a position to help would be due to natural or social circumstance; the question is what we are willing to pay to buy a certain level of protection. As with other insurance schemes, individuals may decide to tolerate different levels of risk. After all, we could decide to take whatever luck brings, neither to give nor ask for help.

In addition to differences in empirical judgment, and the degree of risk people want to tolerate, there are divergent value judgments, as I have noted. How do we rank the various needs which must be satisfied to maintain a basic level of psychophysical functioning or compare any or all of them to injuries? Unless we hold that there is some objective structure of values (of which many are unfortunately ignorant), it seems impossible to avoid the view that there is an ineluctable element of subjectivity in value judgments; there simply is a measure of agreement and disagreement about the relative ranking of various goods. Some agreement is necessary in order to have any principle of mutual aid, e.g., the importance of welfare over against well-being, but we soon come upon disagreement.

Thus for these reasons, variations in empirical predictions, desired level of risk, and value judgments, mutual aid is inherently indeterminate.[23] Perhaps in some human communities agreement is sufficient to achieve a moral consensus or even a legal form of the principle;[24] where the requisite agreement is lacking, individuals settle the meaning of the principle for themselves. There is *agreement* that there be a principle but individuals settle a crucial part of its meaning. The individual must, as it were, legislate for others, although it is really impossible to do so; one acts as one believes others will act. You will risk injury to save my child's life because you hope I would do the same but, in fact, I might not because I don't think you or others would. Lacking a greater degree of agreement, we have to make do in this area of the moral life with an individually determined bond of uncertain and flexible meaning.[25]

Thus in the absence of moral consensus, much less legal definition, we can say that we have an actual convention of mutual aid although the level of cost is indeterminate. According to the sorts of costs we are willing to pay, we would be prepared to yield one or another moral right in order to get a requirement for mutual aid. Would we think of mutual aid itself as a right? Judith Thomson does not want to call her version of mutual aid – 'minimal decency' – a right because she thinks it odd to say that whether one has a right depends on how much it costs others not to violate it ([29], p. 134). (I am not certain whether she even thinks minimal decency is a *requirement*.) The fact, however, that there is a cost clause attached to this requirement would not seem to prevent its being understood as a right since other positive duties of justice may have similar restrictions. What seems to interfere with speaking of mutual aid as a right (which can be demanded or yielded) is its indeterminancy; if I have aided others at a level to which I *hope* they also subscribe, may I demand their aid as my right when the occasion arises? Suppose they have agreed only to a lesser cost. My having given to others at a greater cost would not seem to obligate them to give the same in return to me unless they had agreed to my definition of the principle. Nonetheless, if I *assume* that others have agreed – I believe they have and act accordingly – then I may assume a right to their compliance. It may turn out that I am mistaken; they have no obligation and I have no right to their help. But if my assumption is correct, then they owe the aid and I have a right to it.[26]

III. SUPEREROGATION

Noninjury, duties and rights related to the production and distribution of goods, and the principle of mutual aid are part of the sphere of moral requirement. In addition to what is required, a moral system could extend to that permissible conduct on behalf of others, motivated by altruistic benevolence, which is morally desirable but not required; this we call supererogation.[27] Its desirability consists in the fact that doing this sort of thing (being this sort of person) bolsters, strengthens the web of mutually beneficial social relations. The system of moral requirement provides the floor of social relations; supererogation provides an additional form of behavior which counters egoism and turns us to others. Supererogation supplements the attention to others that justice requires; it flows from our natural concern for others, our altruistic benevolence.

But what does supererogation consist of? For the moment, I will not discuss how one would act in a supererogatory way in the context of the major social institutions; we will have occasion to treat one example of this sort of supererogation when we look at the role relationship in which medical care is delivered. For now let us look at the supererogation which would be contrasted with mutual aid. Let us assume that mutual aid, a duty of reciprocal justice, is defined as I did above: in certain situations, render aid which secures basic human needs, but at no comparable cost. This sort of conduct, of course, could be performed out of altruistic benevolence; the point is that we do not leave it to benevolence but make it required. Over against this notion of mutual aid, there would be two fundamental types of supererogation. Supererogation 1 focuses on the same sorts of situations and the same sorts of aid covered by mutual aid, but removes the limit on the cost to the giver: I will rescue a child from a burning building at a level of danger to myself comparable to or even exceeding that to which the child is exposed. (Defining mutual aid so as to require less from the giver, e.g., a slight cost in well-being, would expand supererogation 1 accordingly.) Supererogation number two focuses not on situations where basic needs are threatened, but on well-being, the remaining range of desires of individuals which may be frustrated by either disasters or the ordinary course of things; I will give you the money to buy a yacht since your salary is low (justly). But this supererogation 2 can take two forms: I will give you the money for a yacht at no comparable cost to myself or I will do so at a comparable or even greater cost to myself. Either form assumes, of course, that the giving is permissible; I have no duty not to use the money in this way or to use it differently. In sum, then, there are two main types of supererogation, one of which has two subtypes:

> *Supererogation 1*: promote welfare, but at a greater cost (comparable or more than comparable) than mutual aid requires.
> *Supererogation 2.1*: promote well-being at no comparable cost.
> *Supererogation 2.2*: promote well-being at a comparable or more than comparable cost.

Note that supererogation 1 has individuals putting the welfare of others above their own; their own welfare is secured through noninjury and principles of justice but they yield or forego their entitlements. They do 'more' than mutual aid requires; in Joel Feinberg's terminology, it is 'duty-plus'.

There are also two other forms of supererogation analogous to 1which we should note at this point. Supererogation 1a covers the situation where a potential *recipient* of mutual aid foregoes, e.g., the right to be rescued. Two of us are in a burning building, but only one of us can be the object of mutual aid. If some just procedure for choosing between us is employed, then we are still within the sphere of moral requirement, but if I say, "I yield my right to be considered, take him or her instead of me," then I have acted out of supererogation 1a. Supererogation 1b covers the case where the *giver* might have been required to render mutual aid, but in fact is not; some just distribution procedure allots the duty to someone else, but the individual volunteers anyway out of benevolence. Thus while supererogation 1 is 'beyond duty' in the sense that one incurs a greater cost than duty requires, one can also promote the good of others by foregoing mutual aid (yielding to others) or by performing it when one is not required to (volunteering).

Supererogation 2.1 involves us in helping others achieve well-being but at no comparable cost; that would mean at no cost in welfare and at no comparable cost in well-being. This is what we often refer to as generosity (cf. [16]). Supererogation 2.2, however, has me helping others achieve well-being but at a comparable and perhaps even greater cost. At a greater cost could mean not only a greater cost in happiness but a cost in welfare as well, e.g., you go without medical care so that I may have a yacht. Note that in this form of supererogation as in the others the costs may be temporary or permanent; they are real, of course, even if reversible. Let us call this self-denial.

To review then, we put mutual aid under the sphere of requirement. It could have been left to benevolence but we justify it in terms of reciprocal justice and make it required. But we limit our aid to that which we can give without causing the same or greater distress for ourselves; helping at a comparable or greater than comparable cost is supererogatory. In regard to well-being, the system of requirement provides for just opportunities. The system corrects the natural and social lottery in ways we deem just; beyond that, we do not agree to require the mutual promotion of well-being. Generosity (2.1) and self-denial (2.2) we leave to those who value giving more than the well-being or the welfare that is justly theirs.[28]

Ought we to do what is supererogatory? The supererogatory is desirable in the sense that the various forms of supererogation counter egoism and bolster regard for others, but it does not follow that we would flatly say that each ought or should be done. Of course, an individual could commit himself or herself to, say, supererogation 1 and assert, "I ought to sacrifice my welfare or life for others; if I am going to live up to my ideal, this is what I ought to do." But may we say others should?

I do not think that there is some special moral sense of ought – albeit not that of requirement – which we use in recommending benevolence.[29] We can say ought to others, but only insofar as we believe that they value benevolence: if you want to be benevolent, then you ought to do what is supererogatory. But parallel to the way in which we advise someone about conflicting moral requirements, we do not simply say without qualification that others ought to perform supererogatory acts. We would generally assume, it seems, that they value their own welfare and well-being and we would advise them accordingly: if you value your own good, you ought not to sacrifice it for others. We may simply bring this to their attention or we may actually let them have our view of which is more important (or our view of what *their* view on reflection would be): "Sure, you want to be benevolent, but you ought not skip a term of college to do famine relief." The value that most of us place on our own self-realization is so great that we are very reluctant to say that even for the sake of benevolence anyone ought to incur a comparable or greater cost, thus we would not ordinarily recommend supererogation 1 (duty-plus) or 2.2 (self-denial). Even to a person who valued benevolence a great deal, we would be reluctant to recommend 2.1 (generosity) where the cost is great; in contrast, where the cost is slight to the giver many would be inclined to say that we ought to be generous.

Do we praise supererogation and blame (dispraise) its omission? We praise it because we value altruistic benevolence but our praise is qualified by our regard for the giver's as well as the recipient's good. Thus we are inclined to praise supererogation 1 (duty-plus) *as* a magnificent beneficence but we would not blame its omission; we might even dispraise it because of the comparable or greater than comparable cost to the giver. From the point of view of *our* valuations, we might condemn it even if we knew that the giver valued self-giving more than self-aggrandizement. Similarly we are inclined to praise 2.2 (self-denial) but not blame its omission; but we are less likely to dispraise it if it only involves a loss of well-being, especially if the giver values this form of supererogation over self-regard. Since we do not (I assume) value well-being as highly as welfare (and don't expect others to) our disapproval of a comparable loss may not mute our praise; it will more likely do so when the cost is very high, e.g., sacrificing a career. As for 2.1 (generosity), here we are inclined to praise and even to blame for omission, since the cost to the giver is not comparable; we are especially inclined to blame (dispraise for lack of benevolence) where the cost to the giver is slight and the gain to the recipient great. In sum, praise and blame in regard to supererogation are forms of admiration and condemnation, approval and disapproval, which express our values, our appreciation of forms of conduct and character: we value both other-regard and self-regard, but some of us rank them differently.[30]

IV. ROLE DUTY

Do mutual aid and supererogatory beneficence have any relevance to medical treatment? Only tangentially, I will argue, for therapeutic personnel have a role specific duty (cf. [16], p. 239).[31]

If there were no specialized class of therapeutic personnel, who practice in and through a particular social institution, then sickness, injury, and other disabilities could be included under the rubric of those situations to which mutual aid applies. But we are accustomed to having health care needs attended to – services rendered – by a special group of providers and organizations. The provision of medical care, like the provision of food, is part of the economy; medical care is one of a number of basic goods and services which we purchase, exchange for value, in a complex division of labor. The provision of medical care extends not only to the attention of medical personnel, but to the distribution of medical supplies such as medicines or organs. Thus unless we encounter a sick or injured person in circumstances where he or she cannot obtain medical attention or supplies or unless it is necessary to help them obtain it (e.g., the neighbor who is too sick to call a doctor), then we do not think that mutual aid applies. In effect, the system of medical care takes the place of mutual aid in regard to this particular need; we distribute medical care through economic structures according to the principles of justice which guide the production and distribution of goods and services.[32]

The relationship which obtains between therapeutic personnel and their clients is conceived in part as an exchange of goods and services; other dimensions of its meaning are provided by moral principles which we believe should structure the exchange.[33] In effect, the exchange is conditional: we agree, but only under conditions X, Y, Z. The exchange and the conditions define the relationship; this is so both for its moral and legal definition, although the two are not equivalent. The therapeutic relation – in which therapists and clients have roles – is complex, but we should note here some of its familiar features:[34] (1) persons who have passed training programs and are properly certified agree to offer their services (as individual practitioners or as part of the staff of health organizations) and in return receive the 'rights and privileges' of the profession; (2) members of the public request their services, pledging payment or having it pledged for them; (3) if the individual in need meets certain conditions (technical and moral), then he or she will be accepted as a client by the therapeutic agent; (4) the relation between the client and the therapeutic agent is defined by a network of duties and rights; some of these will be functionally specific, e.g., therapeutic personnel have the responsibility to perform as well as possible (do good and avoid harm), the client must participate in some way in the treatment; other requirements will be derived from general moral categories which structure the medical relationship, e.g., the right of self-determination, truth-telling, etc.[35]

Bracketing the question whether a relational system such as this should be the mechanism for delivering health care, or, assuming its legitimacy, whether the existing system for the distribution of income and wealth renders the mechanism unjust, issues in medical ethics have primarily to do with the moral and legal definition of the relationship between therapeutic personnel and their clients. I would like to discuss first two matters which are similar to problems we discussed in regard to mutual aid: what costs do therapeutic personnel pledge themselves to incur? Are therapeutic personnel only obligated to treat persons who seek their aid through normal channels?

Let me use a familiar case to get at both questions. Joel Feinberg says the doctor (an example of J. O. Urmson's [31]) who volunteers to go to a plague-ridden city

> is not simply doing his 'duty plus more of the same.' He does not travel a definite number of miles more than the total required by duty; neither does he treat a definite number of patients more than duty requires, for a definite number of hours more than is necessary, at the loss of a definite number of dollars in excess of what is obligatory. The point is, he has no duty to travel one step toward the plague-stricken city or to treat one single victim in it. The whole of his duty as a doctor is to continue treating the patients who constitute his own comfortable and remunerative practice. Still, if he gives all of this up and, at great inconvenience and danger, volunteers to help the suffering in a distant city, his action surely exceeds, in some sense, the requirements of duty ([9], p. 398).

Feinberg draws our attention to two features of the situation; the plague victims are not part of the doctor's ordinary practice; to aid them would involve inconvenience and danger for the doctor.

Now it seems to me that neither of these aspects of the case renders the doctor's action supererogatory. In the first place, consider the problem of risk. If a plague victim appeared in the doctor's office, then the doctor should treat the person. The 1957 Principles of Medical Ethics of the American

Medical Association state that "A physician may choose whom he will serve" but adds "in an emergency, however, he should render service to the best of his ability" ([24], p. 39).[36] Bracketing any other moral consideration, then, it would not seem, even on the AMA's statement, that one of the conditions which permit the doctor not to enter into a contract with a patient is the fact that doing so would impose a comparable or more than comparable risk on the doctor; the degree of risk to his or her health is not relevant. The doctor is not morally obligated to expend personal monies for the individual's treatment – we have bracketed questions about the distribution of income and wealth – but the role requires that the doctor be prepared to take a *health* risk. Over and above specific agreements to provide services to particular persons, therapeutic personnel, we seem to assume, have pledged themselves to offer their aid under conditions which could impose comparable or more than comparable costs. There is a general contract, if you will, between therapists and society, in the context of which particular agreements are made. If it is not just for money that therapeutic personnel pledge themselves to take risks (compare certain construction workers, for example), it is for other rewards. In effect, the rest of us grant them a range of goods in exchange among other things for their willingness to take risks. Thus the agreement to provide medical services is not conditional on there being no comparable risk to therapeutic personnel.[37]

Secondly, the fact that the doctor seeks these patients out would not render the action supererogatory. We would ordinarily expect, as apparently the AMA does, that therapeutic personnel (I am speaking now of a moral definition of their role, what is legally required aside) would respond to an emergency situation in their own environs. For example, suppose there is a massive train wreck; there are too many victims for ordinary emergency care, i.e., ambulances and hospital emergency rooms. Under these circumstances, we would expect therapeutic personnel simply to *go*; those victims who are conscious accept treatment and those who are not competent are assumed to consent just as in ordinary emergency cases. Neither the fact that the therapeutic personnel have to take the initiative – they aren't part of the staff of an emergency room – nor the problem of noncompetence takes the situation outside the moral relation which obtains between therapeutic personnel and their clients. That the plague victims are in some distant city is hardly relevant in itself, assuming that the therapist has no overriding duties at home (to patients or others).

Thus neither the risk nor the mode of the relationship makes the plague situation extraordinary. What distinguishes the Urmson-Feinberg doctor is that he or she steps forward, volunteers, and thus obviates any need for the sort of distribution procedure which is necessary when more than one can help but not all are needed. If the doctor was the only one who could go or if all the available doctors were necessary, then this doctor would have courageously done his or her duty. But in this example we discover a form of supererogation 1b: *voluntarily* taking upon oneself a task which a just distribution procedure might not have imposed. It is not an action which the doctor would never be *obliged* to perform under any circumstances, for under certain conditions it would be required; nor is it 'duty-plus' in the sense of suffering a greater risk or loss (more hours, more danger) than required; it is doing more than his or her share in the special sense that a just distribution procedure might not have required *this* doctor to go at all.

If we changed the case, however, the doctor's going could be interpreted as mutual aid. Suppose that neither through the plague victims, nor their relatives, nor the government of that 'distant city', nor through any social mechanism was the doctor to be compensated; no *system* of compensation applies to this case. In these circumstances, we could think of the doctor acting from mutual aid at whatever level of cost his or her principle calls for. The doctor would not be acting *qua* therapist, that is, in the therapeutic role, but as an individual who happens to have special skills. The doctor would be like a lifeguard who rushes to help even when those about to drown are not part of his or her ordinary contractual clientele and when there is no social mechanism to provide compensation. The special skills are relevant to the distributive question of who should provide help in a particular situation, just as a firefighter's or mountain climber's might be (we assume that if more than one can help, the duty falls on whoever, *ceteris paribus*, can do so most efficiently). The lifeguard does not receive specific compensation but is paid in the coin of the adherence of others to the principle of mutual aid.

In contrast to this view of the situation, where the individual acts outside of his or her role, therapeutic personnel (or lifeguards) could even stipulate as *part* of their role definition that in some circumstances services should be rendered when ordinary compensation is not available. Role definitions are sufficiently fluid to permit this move; the inclusion of *pro bono* work is justified as an exchange for other socially conferred benefits. If the aid is conceived as outside the role, as mutual aid, then the therapist would stop short of helping at a comparable or greater than comparable cost; going to these levels would move the action into the category of supererogation

1 (duty-plus).[38] But if the noncompensated aid is rendered as part of the role, then presumably the therapist would be prepared to take the same risks as in any ordinary case.

What forms of supererogation appear in the medical context? Are there analogues to supererogation 2.1 (generosity) and 2.2 (self-denial)? I heard recently of a teenager who had cystic fibrosis. At one point in his illness, which was eventually fatal, he was asked whether he wouldn't like to go to a hospital in his own city rather than to go back to a famous children's hospital some distance away. No, said the boy, he would rather go to the children's hospital; it's like home there.

Now what could he have meant? He could have been saying that the children's hospital was familiar, that he was *accustomed* to it; perhaps he meant that he has *friends* there, other patients and staff; he could have been saying that the staff there do what their roles require out of *benevolence*, not merely from duty. No doubt he meant some or all of these things, but it is likely that he was also saying this: the staff at the hospital do not merely perform their role duty, but something more. They exhibit a concern not only for his welfare but for his well-being and they promote it, some of them perhaps even at a comparable or more than comparable cost to themselves. It is difficult to distinguish, of course, between the role requirement to care for the psycho-physical welfare of the client and a supererogatory devotion to his well-being. If the patient's welfare cannot be treated without an attention to his entire outlook, given the variety of psychological factors which influence it, then it is hard to separate the psychological care which therapeutic personnel are required to provide in their role capacities from a broader concern for the client's well-being. Nonetheless, in this case I think it likely that what was done exceeded role requirements.

Evidently this boy gladly received supererogatory attention and it is natural that he would. But supererogation 2.1 (generosity) and 2.2 (self-denial) can conflict with the ideal of self-sufficiency. Some people would not welcome the supererogatory efforts of medical staff; we sometimes feel that the kindly attentions of others are demeaning and we reject their gifts (with polite appreciation, if we excuse their ignorance of our desires). We respect the right to refuse treatment or mutual aid, and the right to self-determination would also seem to require that we not impose our generosity or self-denial.[39] Adults as well as children, of course, often receive supererogatory benevolence gladly; they might even be prepared to sacrifice their own desire for self-sufficiency in order not to hurt the feelings of the misguided giver, thus performing a supererogatory act of their own. These relations are complex; the tension between receiving and standing alone will be resolved according to the sorts of persons we are, the ideals that move us.

V. CONCLUSION

Against an assumed background of other duties and rights, I have presented mutual aid as an actual convention; we yield entitlements in order to gain protection. Since our agreement with others is motivated by self-interest, it stands in contrast to altruistic benevolence; it also seems a pale substitute for the notion that our duty to aid others is based on the inherent requirements of reason and agency. The actual convention account, however, is the most satisfactory one I know of; it explains the divergence between those who feel that human fortune and misfortune should be shared, even equalized, and those who feel that while some minimal aid should be rendered, we should not tamper further with the lottery of events, at least in the restricted sorts of situations to which mutual aid applies. The indeterminancy of the convention also explains the uncertainty of the line between mutual aid and supererogation. As for medical care, I have treated it as part of the economic system; the need it addresses is not left to the ordinary convention of mutual aid but is put under the umbrella of the principles and institutions in terms of which goods and services are distributed. Many of the forms of supererogation can be (and often are) present in the medical context, but the beneficence which is proper to the therapeutic role is a duty of justice.

I have presented a picture, of course, of individuals the majority of whom are motivated more by self-interest than benevolence. This is not a picture which simply mirrors capitalism or 'economic rationality', but it is rooted in a view of human nature, or at least of the human condition as we now experience it.[40] Requirement – mutual aid or roles – is primary, benevolence only supportive.[41]

Brown University
Providence, Rhode Island

NOTES

[1] Cf. ([4], p. 139). To propose beneficence, understood as a type of utilitarianism, as the principle of justice for social institutions would be a different matter; I do not deal with this issue here.

[2] It would be interesting to investigate historically the relation of benevolence (and the compassion from which it springs) and requirement in various moral systems.

[3] For a review of interpretations, see ([12], pp. 39–45).

[4] Cf. treatment of this case and Kitty Genovese in [29].

[5] See [5].

[6] If my action puts others at risk outside the context of consensual social practices, then there might be a duty above and beyond mutual aid. However, see [29] on the case of abortion.

[7] Cf. ([19], p. 38; [13], p. 54; and [22], p. 55).

[8] For a different notion of the sorts of aid one is required to give, compare the principles of beneficence suggested by Beauchamp and Childress, "the duty to help others further their important and legitimate interests when we can do so with minimal risks to ourselves." This broad category of interests would seem to extend beyond welfare to well-being; we have a duty "to be as beneficent as possible within our means." Beauchamp and Childress later restrict aid to situations where someone is "at risk of significant loss or damages" ([4], pp. 136–140).

Singer begins his well-known article by speaking of "suffering and death from lack of food, clothing, shelter, and medical care," but his principle refers broadly to "preventing something bad" ([27], p. 24).

[9] Michael Slote distinguishes between preventing suffering and positive acts of beneficence ([28], pp. 125–127). Cf. ([27], p. 24). I assume that by the prevention of suffering he means stopping it from occurring and causing it to cease. He seems to conceive of suffering as a state of affairs and a correlated affective response; it seems close to what I refer to as deprivations of life, limb, or basic needs. His notion of 'beneficence', in contrast, seems to refer to the promotion of well-being.

[10] William Aiken would have us save others from death by deprivation unless we would expose ourselves to the same danger or incur an "unreasonable" cost, a standard he leaves "flexible". Aiken thinks that the stringency or strength of the duty lessens as the cost to the giver increases, but I am uncertain what stringency means; it would seem that one either has a duty or one doesn't, the cost either falls within the proper range or not ([1], pp. 95–96).

Cf. John Arthur who does not require us to prevent the "death of an innocent" if we would have to sacrifice "anything of substantial significance" ([3], p. 47). He means both other duties and one's own 'long-term happiness'.

[11] See ([28], p. 125–128).

[12] John Kleinig suggests that we should set aside the pursuit of our non-welfare interests to aid others; we should sacrifice our welfare interests only if the loss is temporary. He would not have us put aside our nonwelfare interests up to the point of comparable cost, however. First, to help in one situation does not require us to help in morally similar situations for that would require of us "an impossible moral herculeanism." Second, it would destroy our autonomy, for, according to circumstances, "nearly all I have and hold dear will need to be sacrificed." It is not clear what Kleinig means by "impossible"; certainly it is possible to give up to the level of comparable cost either on one occasion or over several. Moreover, it might be that we judge it our moral duty to suspend our own engagement in "valuable pursuits"; we would not lose *moral* autonomy, only the pursuit of our nonwelfare interests. It does not seem to me that this is a question of the "value" of our pursuit *per se*, of the "hierarchy of our interests," or of some particular

good (e.g., Mary's anointing the feet of Jesus). It is a question of the level of costs we are required to sustain. Values themselves do not override requirements, although the valuation we set on our own pursuits is a factor in deciding which requirement to accept ([17], pp. 385, 396–397).

13 Slote [28] also discusses "commissive" immoralities in the use, retention, or increase of wealth, e.g., keeping servants, having money that was wrongly acquired by you or your benefactors, investing in morally flawed enterprises. While he argues against a costly principle of aid, Slote believes that much of the world's wealth ought to be redistributed given past and present injustices. One could also argue that the present *system* is unjust, root and branch.

14 See ([3], p. 40).

15 William Aiken points out that the giver must be aware of the need and have the means to satisfy it ([1], pp. 91ff). Having signifies moral and legal access, but there are exceptions: one might be justified in taking someone's unjustly acquired goods in order to save others or one might be justified in appropriating goods from someone who neglects the duty to save.

16 See [27] and [5]. As William Aiken rightly notes, the fact that others can help doesn't relieve me of the *prima facie* duty to do so. Those who can help have a *prima facie* duty to do so; whether it is actual, their duty overall, would depend on how the burden is distributed between givers and on other duties the giver may have. See Aiken on being the "last resort" and on conflicting duties ([1], pp. 93–98).

17 Alan Gewirth grounds mutual aid, in neo-Kantian fashion, on the nature of human agency and the principle of universalizability: " . . . whenever some person knows that unless he acts in certain ways other persons will suffer basic harms, and he is proximately able to act in these ways with no comparable cost to himself, it is his moral duty to act to prevent these harms . . . By 'comparable cost' is meant that he is not required to risk his own life or other basic goods in order to save another person's life or other basic goods . . . To engage in such a risk or to incur such cost would involve the possibility or actuality of losing his own life in order to save theirs, and this, rather than maintaining an equality of generic rights, would generate an inequality in his recipient's favor . . . " ([13], pp. 217–218). Gewirth's 'duty to rescue' is more than a principle of mutual aid in my sense; it is the basic positive duty which regulates social institutions as well as the rendering of aid in special situations.

18 Note that it is not simply because a cost limit is set that mutual aid is part of the system of moral requirement and not an expression of altruistic benevolence. The altruistic giver can also set a limit to the costs he or she is willing to incur, but within those limits the giver is prepared to aid others regardless of whether any good accrues to the self from the interaction. For example, I might not be prepared to give my life for a drowning swimmer (assume I don't have any special role duties), but out of altruistic benevolence I make the attempt at some lesser cost regardless of whether I am rewarded or any other good accrues to me. If I had acted on the basis of mutual aid, however, I would have expected a return in the coin of the subscription of others to the same principle.

19 Cf. [32], pp. 144–149; also cf. D. A. J. Richard's Rawlsian justification of mutual aid as a principle rational contractors would agree to from an 'original position' of moral equality. Rightly insisting on a distinction between mutual aid and 'general beneficence', Richards restricts the principle of mutual aid to cases where great pain, injury and death

are at stake and specifies that aid is to be rendered only at a 'slight cost' to the giver. His argument is that the contractors, following a maximin policy, would not agree to sacrifice life and limb to aid others: they would be agreeing to risk their lives, he says, "where they are not likely to be successful and lose their own lives in the process" ([25], p. 187). But not all situations are ones in which aid is not likely to be successful. Would the contractors agree to a more costly requirement where the likelihood of success is great? Would the contractors agree to give their lives even if the outcome were certain? Why wouldn't they agree to give aid up to the point of comparable loss? This would promote welfare for the unlucky but would not require givers to lose their lives. For an account of this sort, see ([22], pp. 59ff): " ... you get the protection implicit in knowing that so long as there is enough for everybody, you will at least get enough to keep you going."

20 See [5].

21 Cf. ([15], p. 105; and [5]).

22 Harman defends his theory at more length in [14]. His thesis is that "inner judgments" – some person ought or ought not to do X, was right or wrong to have done or not done X – assume a set of shared reasons or motivations, in particular a set of shared agreements. I gather he means that we should not make inner judgments about Hitler for that would assume that he shared our agreements but that we could condemn Hitler's acts from the perspective of agreements we share with others. It was wrong to kill the Jews and it would have been wrong *of him* had he shared our agreements ([15], p. 7). It would not be necessary, of course, to explain the moral principles on the basis of which we make inner and other judgments as agreements, as actual conventions; this is a separable part of Harman's thesis.

23 Slote [28] raises the question whether a costly standard of aid would be too difficult for most people; failing to do right in this regard, they might relax their adherence to all moral rules; see Singer's reply ([27], p. 29). Despite the unlikelihood of these consequences, we would have to consider other possible results. As Brock [5] notes, at what point would the sacrifice required by mutual aid inhibit accumulation?

24 An additional question is whether mutual aid ought to be legally defined and enforced. One strand of argument claims that to omit aid is the cause or at least a causal factor in prolonging or exacerbating harm; since the law concerns itself with harm, a Good Samaritan principle, *ceteris paribus*, should be legalized [17]. The central objection to this argument is that omitting aid is not a cause or a causal factor, strictly speaking. Speaking of it in causal terms masks a normative position – aiding others should be legally defined and enforced [20]. As Mack notes, a number of legal authors make the argument for legalization independent of the causation thesis. My view is that even if omitting to help were a cause or causal factor related to harm, it is morally distinct from the basic prohibition against harming others and from specific role responsibilities which reflect our principles of social justice (cf. [17]); its legalization therefore also requires independent justification.

25 Harman ([14], pp. 15ff) tries to refute traditional objections to agreement theories. In particular, to questions such as when did we agree and what about those who don't, he replies that by an 'agreement' he is talking about having intentions on the assumption that others have them; there is no one moment at which an explicit agreement is made. One's intentions would provide presumably for how to deal with those who remain outside the convention ([14], p. 19).

26 William Aiken characterizes the right to be saved from death by deprivation as a right in a 'normative' but not a 'descriptive' sense ([1], p. 89). He means that at present there is no generally agreed upon moral or legal rule which protects this right ([1], pp. 87ff). I assume, however, that there is enough moral agreement about mutual aid, despite its indeterminancy, to think of it as a right descriptively as well as normatively. Furthermore, since a descriptive right in Aiken's sense is normative as well – it expresses a justified moral rule or judgment – perhaps it would be better simply to distinguish those rules which have general moral acceptance (and perhaps legal form) from those which don't; presumably there could be a spectrum of agreement and disagreement ranging from commonly accepted principles to radical criticism.
27 See Schumaker's concise formulation: "'supererogation' = df. acts the performance of which are conjunctively (1) neither forbidden nor required by morality, (2) good, and (3) done primarily for the sake of someone other than the agent" ([26], p. 11). Schumaker's essay is a rich treatment of the meaning and the varieties of supererogation; he contrasts supererogation with the sphere of rights, duties, and justice ([26], pp. 30–35). For a discussion of the logic of requirement and supererogation see [7]. (I do not discuss here Chisholm's category of offense which is the parallel to prohibition.)
28 See [6] and [25]; cf., [16] and [32].
29 W. K. Frankena distinguishes between (1) "What is strictly required or obligatory . . . ; (2) What is strictly forbidden or wrong . . . ; (3) What ought to be done . . . but is not . . . a duty or obligation . . . ; (4) What ought . . . not to be done but is not strictly forbidden or wrong . . . ; (5) What is indifferent . . . " ([10], p. 104). (His scheme is similar to Chisholm's five-fold classification.) Frankena says that (3) and (4) are part of a moral requirement, not just morally good or bad ([10], pp. 103–104). On my view, however, the ought in (3) and (4) is not part of the system of requirement but expresses our valuations. Moreover, it seems to me that we would want our principles of justice to be strictly required. Frankena, in contrast, says that although we ought to do good (benevolence) and treat people equally (justice), morality does not strictly insist on either ([10], p. 106).
30 Richards sees supererogation as required (with the exception of 'beneficence', which can involve the sacrifice of the interests, even the life, of the giver), but not as a duty ([25], p. 197); for Richards anything that is a duty is subject to justifiable coercion. In contrast, in my view none of the varieties of supererogation are requirements; they express the benevolence of the giver. The blame we express for the failure to do supererogatory acts is not therefore an analogue or substitute for legal enforcement, as it is for Richards ([25], p. 198); it is a function of our valuation of altruistic benevolence.
31 'Medical personnel', 'therapeutic personnel', and 'therapist' are all awkward expressions, but they are the best I can do. 'Doctor' won't work, of course, because nurses and others provide treatment. 'Health care professional' is too broad because that would include various auxiliaries such as administrators and lab technicians; 'therapeutic agent' is misleading since it gives the impression that one party is active and the other is only the 'patient', the one who receives, who suffers the action of another.
32 The production and distribution of health services has many dimensions, of course, e.g., preventive or public health measures. Note also that some provisions for the disabled are not part of health services per se, but of arrangements for the just distribution of social opportunities. Cf. ([14], pp. 136–137).

[33] John Ladd ([18], esp. pp. 22–27) argues that in addition to rights and supererogation, morality includes relationships in which persons, e.g., friends, doctors/nurses, parents, have the responsibility to promote the long-term well-being of others. A contractual or role notion of rights and duties is inappropriate for these relationships; one responds to needs out of love or concern; the standard is to each according to his needs, from each according to his ability. While moral rights do apply in the medical context, the relationship of responsibility is primary. My role model, understood in the context of medicine as an institution, is contrary to Ladd's of course. On my view, the role model can account for some of the features Ladd sees in responsibility relationships, e.g., attitudes and motives ([18], p. 28). It could not include "from each ... to each" as a standard since it envisages a contractual exchange.
[34] For a more detailed account of role morality and the clinical context, see ([30], pp. 337–346).
[35] The client will *trust* therapeutic personnel for the execution of policy; the client may or may not delegate decisions about strategy or tactics.
[36] See the other selections on the physician-patient relationship in Part I of [24], in particular Edelstein's discussion of the Hippocratic *Precepts*. This doctrine was reaffirmed by the AMA in its 1980 statement.
[37] I assume here only that the therapist is obliged to provide emergency services. George Annas suggests the broader thesis that physicians have a moral, and should have a legal, duty to render not just emergency (immediate threats to life or health) but 'essential' services ([2], pp. 16–17). Essential services are those where a delay in treatment would mean more extensive or expensive care, or substantial pain and discomfort. My characterization of the therapeutic role and Annas's claim both oppose the notion that the therapist may 'accept' whomever he or she pleases for whatever reason. One must set, however, the technical and moral conditions under which a therapist may morally or legally refuse a patient.
[38] See William May's critique of codes which suggest that physicians accept their duties out of generosity ([22], pp. 70–71). But May argues that it would be wrong to reduce the physician-patient relation to a contract; he insists that it should have a 'donative' element so that the doctors will go the 'second mile' ([22], p. 73). My view, in contrast, is that a type of medical mutual aid may or may not be written into the role; supererogatory beneficence is external but supportive.
[39] Richards adds to his principle of mutual aid the proviso that the putative recipient "does not voluntarily and rationally refuse it" ([27], p. 187).
[40] If human nature or the conditions of existence changed, individuals might reject the entire system of moral requirement. See [23].
[41] A number of people have commented on and criticized this paper; I would like especially to thank Dan Brock and Paul Lauritzen.

BIBLIOGRAPHY

1. Aiken, W.: 1977, 'The Right to be Saved From Starvation', in W. Aiken and H. LaFollette (eds.), *World Hunger and Moral Obligation*, Prentice-Hall, Inc., Englewood Cliffs, N.J., pp. 85–102.
2. Annas, G.: 1978, 'Beyond the Good Samaritan: Should Doctors Be Required to Provide Essential Services?,' *Hastings Center Report* 8 (April), 16–17.

3. Arthur, J.: 1977, 'Rights and the Duty to Bring Aid', in W. Aiken and H. LaFollette (eds.), *World Hunger and Moral Obligations*, Prentice-Hall, Inc., Englewood Cliffs, N.J., pp. 37–48.
4. Beauchamp, T. L. and Childress, J. F.: 1979, *Principles of Biomedical Ethics*, Oxford University Press, New York and Oxford.
5. Brock, D.: 1982, 'Utilitarianism and Aiding Others', in H. Miller and W. Williams (eds.), *The Limits of Utilitarianism*, University of Minnesota Press, Minneapolis, in press.
6. Carney, F. S.: 1973, 'Accountability in Christian Morality', *Journal of Religion* **53**, 309–329.
7. Chisholm, R. M.: 1968, 'Supererogation and Offense: A Conceptual Scheme for Ethics', in J. J. Thomson and G. Dworkin (eds.), *Ethics*, Massachusetts Institute of Technology, Cambridge, pp. 412–429.
8. Edelstein, L.: 1977, 'From "The Professional Ethics of the Greek Physician",' in S. J. Reiser, *et al.* (eds.), *Ethics in Medicine: Historical Perspectives and Contemporary Concerns*, The MIT Press, Cambridge and London, pp. 40–51.
9. Feinberg, J.: 1968, 'Supererogation and Rules', in J. J. Thomson and G. Dworkin (eds.), *Ethics*, Massachusetts Institute of Technology, Cambridge, pp. 391–411.
10. Frankena, W. K.: 1970, 'The Principles and Categories of Morality', in J. E. Smith (ed.), *Contemporary American Philosophy*, George Allen and Unwin, Ltd., London, pp. 93–106.
11. Frankena, W. K.: 1973, *Ethics*, Prentice-Hall, Inc., Englewood Cliffs, N.J.
12. Furnish, V.: 1972, *The Love Command in the New Testament*, Abingdon Press, Nashville, TN.
13. Gewirth, A.: 1978, *Reason and Morality*, University of Chicago Press, Chicago.
14. Harman, G.: 1975, 'Moral Relativism Defended', *Philosophical Review* **74**, 3–22.
15. Harman, G.: 1977, *The Nature of Morality*, Oxford University Press, New York.
16. Hunt, L.: 1975, 'Generosity', *American Philosophical Quarterly* **12** (July), 235–244.
17. Kleinig, J.: 1976, 'Good Samaritanism', *Philosophy and Public Affairs* **5** (Summer), 382–407.
18. Ladd, J.: 1978, 'Legalism and Medical Ethics', in J. W. Davis, *et al.* (eds.), *Contemporary Issues in Bioethics*, The Humana Press, Clifton, N.J., pp. 1–35.
19. Little, D. and Twiss, S. B.: 1978, *Comparative Religious Ethics*, Harper and Row, San Francisco.
20. Mack, E.: 1980, 'Bad Samaritanism and the Causation of Harm', *Philosophy and Public Affairs* **9** (Spring), 230–259.
21. May, W. F.: 1977, 'Code and Covenant or Philanthropy and Contract?', in S. J. Reiser, *et al.* (eds.), *Ethics in Medicine: Historical Perspectives and Contemporary Concerns*, The MIT Press, Cambridge and London, pp. 65–78.
22. Narveson, J.: 1977, 'Morality and Starvation', in W. Aiken and H. LaFollette (eds.), *World Hunger and Moral Obligation*, Prentice-Hall, Englewood Cliffs, N.J., pp. 49–65.
23. Reeder, J. P., Jr.: 1980, 'Assenting to Agape', *Journal of Religion* **60** (January), 17–31.
24. Reiser, S. J., *et al.*, (eds.): 1977, *Ethics in Medicine: Historical Perspectives and Contemporary Concerns*, The MIT Press, Cambridge and London.

25. Richards, D. A. J.: 1971, *A Theory of Reasons for Action*, Clarendon Press, Oxford.
26. Schumaker, M.: 1977, *Supererogation: An Analysis and Bibliography*, St. Stephen's College, Edmonton, Alberta.
27. Singer, P.: 1977, 'Famine, Affluence, and Morality', in W. Aiken and H. LaFollette (eds.), *World Hunger and Moral Obligation*, Prentice-Hall, Inc., Englewood Cliffs, N.J., pp. 22–36.
28. Slote, M.: 1977, 'The Morality of Wealth', in W. Aiken and H. LaFollette (eds.), *World Hunger and Moral Obligation*, Prentice-Hall, Inc., pp. 124–147.
29. Thomson, J.: 1973, 'A Defense of Abortion', in J. Feinberg (ed.), *The Problem of Abortion*, Wadsworth Publ. Co., Belmont, California, pp. 121–139.
30. Twiss, S. B., Jr.: 1977, 'The Problem of Moral Responsibility in Medicine', *The Journal of Medicine and Philosophy* 2, 330–375.
31. Urmson, J. O.: 1958, 'Saints and Heroes', in A. I. Melden (ed.), *Essays in Moral Philosophy*, University of Washington Press, Seattle and London, pp. 198–216.
32. Wallace, James D.: 1978, *Virtues and Vices*, Cornell University Press, Ithaca and London.

SECTION II

BENEFICENCE IN RELIGIOUS ETHICS

RONALD M. GREEN

JEWISH ETHICS AND BENEFICENCE

Medical care and illness have always provided opportunities for the exercise of altruism and beneficence, but in our own day these opportunities have in some ways expanded. The growing field of medical experimentation and the possibility of organ transplantation and donation, for example, have provided new occasions for acts of selfless devotion to the welfare of other persons. Against the background of these very contemporary issues, it is useful to ask what the attitude of a traditional faith like Judaism is to acts of beneficence. Does traditional Jewish teaching, as elaborated in major legal texts and rabbinic writings, permit persons to subordinate their interests in order to promote the welfare of others? Does Judaism ever require an individual actively to risk his interests or his life on others' behalf? And what are the implications of answers to these questions for one's participation in experimentation, organ donation, or other forms of medical beneficence? Because of the richness of the Jewish tradition of moral reflection, the answers to these questions are of interest not only to those who identify themselves as Jews but also to those who seek to come to terms with the difficult new choices imposed by modern medical care.

In the contemporary period, orthodox Judaism's estimate of acts of self-sacrificial beneficence has actually been a source of considerable debate within the Jewish intellectual community. Confronting Christianity's demanding ethic of sacrificial love as epitomized by the teaching "Greater love hath no man than this, that a man lay down his life for his friends" (John 15: 13), Jewish thinkers have been challenged to identify their own faith's attitude to extreme acts of beneficence. Some have sought to defend the distinctiveness of Jewish teaching. They have maintained that Judaism, with its stress on the worth of each individual, has no place for dangerous acts of heroism. Other writers have rejected these claims, however, and have argued that on the matter of self-sacrificial love for others, Judaism and Christianity are in substantial agreement. At the heart of much of this controversy is a Talmudic text found in the tractate *Baba Meẓi'a*. In the course of developing the meaning of the Biblical verse "*that thy brother may live* with *thee*" (Leviticus 25: 36), the text presents a hypothetical case involving a classic conflict of life with life:

Earl E. Shelp (ed.), Beneficence and Health Care, 109–125.

If two are travelling on a journey [far from civilisation] and one has a pitcher of water, if both drink, they will [both] die, but if one only drinks, he can reach civilisation, – The Son of Patura taught: It is better that both should drink and die, rather than that one should behold his companion's death. Until R. Akiba came and taught: '*that thy brother may live* with *thee*:' thy life takes precedence over his life (62a).[1]

Commenting on this text some decades ago, the great Zionist spokesman Ahad Ha'am saw it as expressing Judaism's unique moral genius. He concedes that Judaism appears to reject the extreme altruism of the Gospels, but he dismisses Christian altruism itself as morally questionable:

... [T]he 'altruism' of the Gospels is nothing but an inverted 'egoism.' It strips man of his *objective* moral worth, as it pertains to himself, and makes him the means for a subjective aim. But whereas 'egoism' makes the other into a means for the advantage of the 'self,' 'altruism' makes the 'self' into a means for the advantage of the other.

Judaism, however, removes from morality the subjective relationship, and establishes it on an *abstract objective* basis: on *absolute justice* which sees man as a substantial moral value, without distinction between the 'self' and the 'other' ... Just as I have no right to destroy the life of the 'other' for the sake of the life of the 'self,' so I have no right to destroy the life of the 'self' for the sake of the life of the 'other' ([16], pp. 107f).

For Ahad Ha'am, therefore, and for other Jewish scholars who share his view,[2] the strict impartiality and fairness of traditional Jewish ethics and law is evidenced by this instance of repudiation of extreme beneficence.

In a series of articles and books over the past few decades Louis Jacobs has defended a far different view both of this text and of Judaism's attitude toward self-sacrificial conduct.[3] According to Jacobs, the long history of Jewish martyrdom bears ample testimony to the belief that it is acceptable and even noble to give one's life on others' behalf ([12], p. 45). He questions Ahad Ha'am's reading of the *Baba Meẓi'a* passage and, on independent grounds, speculates that it might be the altruistic position of the Son of Patura, not the ordinarily respected view of Akiba that should be here regarded as authoritative. He further insists that in any case, the principal lesson of this text is not that one's life and interests take precedence over those of the neighbor, but rather that if the choice is between allowing at least one person to live or sharing the water and *both* dying, the choice that saves a life should be made ([12], p. 46). Jacobs concludes by affirming that despite the difficulties it presents, the ethical possibility of self-sacrificial beneficence is firmly rooted in the Jewish tradition:

... of course ... it would be an impossibility for *all* men to attempt to go through life, loving their neighbors *more* than themselves. *As thyself* is the only rule for society.

But the rare individual, who in a moment of tremendous crisis, can rise to the heights of giving his life for his friend – like the Sidney Carton of fiction and the Captain Oates of fact, is a saint and would be recognized as such by Judaism. Jewish history has not lacked such 'Fools of God' ([12], p. 47).

Clearly we are dealing here with a controversy in which apologetic motives color presentations on both sides of the issue. Yet it is also true that the classical Jewish estimate of self-sacrifice, beneficence and altruism is so complex that it admits a variety of different interpretations. In what follows I propose to sort out some of the different strands of Jewish teaching on this matter, and I also want to explore the implications of this teaching for contemporary issues in medical ethics. In pursuing this task, however, I am keenly aware of the fact that there is a great diversity of teaching within Judaism on almost every subject of legal or moral importance. This makes speaking of a single 'normative' Jewish view very difficult. Nevertheless, there are dominant perspectives within the tradition and positions on which the majority of scholars appear to agree. When I speak of the 'Jewish' view of an issue, therefore, I necessarily imply only that this is a view with wide, but rarely unanimous, support in the community of scholars.

II

As we look at classical Jewish teaching on beneficence, it is perhaps important at the outset to distinguish between two different kinds of active, other-regarding conduct: that which risks the interests or material possessions of an individual but poses no threat to his life and that which exposes his life to jeopardy. Obviously this distinction is not a neat one, but it is important to bear it in mind since traditional Jewish thinkers appear far more willing to permit or to encourage beneficence in cases where the agent's life is not jeopardized by his helpful act than they are in cases which place his life in peril. Thus, Jewish teaching is rife with injunctions to "*love thy neighbor as thyself*" (Leviticus 19: 18) and Jewish religious law (*halakhah*) sometimes enjoins the individual to promote the neighbor's welfare when lesser issues than life are involved. The doing of deeds of loving kindness (*Gemillath Hassadim*), for example, is a requirement binding on the pious Jew and extends to fulfilling the neighbor's needs, whether material or psychological. This requirement underlay the elaborate system of charity developed in traditional Jewish communities, and it also helped inspire the delicate structure of norms encouraging respect for the dignity and self-esteem of the under-privileged or destitute individual.[4] The spirit of *Gemillath Hassadim* is expressed in the description the Rabbis offer of the *Am-haareẓ* (the poor) and the *Hasid* (the genuinely pious man). The former says, 'What is mine is yours and what is yours is mine" but the latter says "Mine is yours and what is yours is yours" ([6], *Aboth*, 5: 10). The *Hasid* refuses to place sharing on a basis of equal entitlement and is willing to give without expectation of reciprocity.

Despite this very strong encouragement to sharing and the requirement of charity, the Rabbis endeavored to place limits on the scope of generosity even when no mortal risk to the self was involved. They appear to have believed that there is a principle of personal responsibility to oneself, a principle grounded in the reasonable idea that each individual is his own best protector. As a result, they prohibited individuals who sought to aid others from doing so to the extent that they became themselves objects of the community's charity. The consistent Rabbinic teaching was that during his life an individual was not permitted to give more than twenty percent of his resources to charity ([6], *Ketuboth* 50a). Following his death, when presumably this issue of self-support vanished, this restriction was eased ([6], *Ketuboth* 67b).

This same insistence on personal responsibility and the positive duty to aid others underlies the elaborate system of priorities for lending established by the Rabbis. These priorities are succinctly stated in a passage from the tractate *Baba Meẓi'a* commenting on the verse "*If thou lend money to any of my people that is poor by thee*" (Exodus 22: 24):

> [this teaches, if the choice lies between] my people and a heathen, '*my people*' has preference; the poor or the rich – the '*poor*' takes precedence; thy poor [sc. thy relatives] and the [general] poor of the town – thy poor come first; the poor of thy city and the poor of another town – the poor of thine own town have prior rights . . . The Master said: '[If the choice lies between] my people and a heathen – '*my people*' has preference.' But is it not obvious? – R Nahman answered: Huna told me it means that even if [money is lent] to the heathen on interest, and to the Israelite without [the latter should take precedence] ([6], *Baba Meẓi'a* 71a).

The close of this passage indicates that we are not here dealing with any kind of concession to egoism or selfishness, for if that were the case profitable assistance to the stranger would surely be allowed. Rather, we are dealing with a moral principle that places responsibility for the maintenance of each individual or community on that individual or community. The Rabbis, it seems, were suspicious of a different approach that would place primary responsibility for the self on others, and they possessed the confidence that a strong principle enjoining each to look after his own was the firmest basis for individual protection and communal flourishing.

Where beneficence involving material possessions or less-than-vital goods is concerned, therefore, the Rabbis elaborated a position which encouraged, and to a degree even required, qualified self-sacrifice. Where human life was involved they extended this thinking although, as we shall see, the qualifications grew even more rigorous. Underlying all Jewish teaching in this area is the extraordinary value Judaism places on human life. This is vividly illustrated in Mishnah *Sanhedrin* (the original legal core of the Talmudic tractate by this name). In connection with the admonition to witnesses in capital cases, the text expounds the meaning of the verse "*the bloods of thy brother cry*" (Genesis 4: 10):

It says not 'the blood of thy brother,' but *The bloods of thy brother* – his blood and the blood of his posterity ... Therefore but a single man was created in the world, to teach that if any man has caused a single soul to perish from Israel Scripture imputes it to him as though he had caused a whole world to perish; and if any man saves alive a single soul from Israel Scripture imputes it to him as though he had saved alive a whole world ([4], Mishnah *Sanhedrin*, 4: 5).[5]

Obviously, this teaching has ambiguous meaning for the issue of beneficence and altruistic self-sacrifice. It can strongly support sacrificial acts where the saving of life is involved, but it can also prohibit these where beneficence exposes the self to danger. The Rabbis appear to have developed both these implications.

The very high priority given the duty to save another's life is well illustrated by the classical Jewish insistence that virtually any of the commandments – with the exception of the prohibitions against murder, incest or adultery, and idolatry – may be violated to save a human life ([6], *Sanhedrin*, 74a). This teaching was grounded in an interpretation of Leviticus 18: 5 "*that man shall live by them*" which was taken to mean that the divine commandments have life, not death, as their object. In this connection, a specific principle, *Pikku'ah Nefesh* (the saving of a soul) was elaborated to permit all sorts of sacramental violations when risk to life was involved. Thus, despite its sanctity the sabbath might be transgressed to save someone whose life is in danger or to succor a seriously ill person ([6], *Shabbath*, 150a; *Yoma* 84b), and the Rabbi of a community where this fact was not widely known was held to be seriously derelict in his duty as a teacher. Indeed, going beyond this permission for sacramental violations, it was the view of great scholars like Moses Maimonides and Joseph Caro that anyone who sacrifices his life to fulfill a religious precept when this is not explicitly required by law is guilty of committing a deadly sin ([13], p. 53).

It follows from this stress on the value of human life that at least where no serious personal risk is involved, every individual is duty-bound to rescue or to aid an individual whose life is in danger. According to Rabbinic teaching, anyone refusing to aid a person in danger of losing life, limb or property is guilty of violating the specific Biblical commandment "*Thou shalt not stand idly by the blood of thy neighbor*" (Leviticus 19: 16).[6] This obligation was held to be especially incumbent on the physician. Far in advance of contemporary 'good Samaritan' laws, Jewish teaching stressed the physician's positive obligation to extend medical assistance. Failure to do so, unless a more competent physician is available, is regarded as tantamount to bloodshed and as subject to punishment ([3], 336: 1). At the same time, this duty to save a life was not limited to the committed professional but extended to any individual able to be of assistance. It is true that no specific religious-legal (*halaknic*) penalties were indicated for the private individual who failed to come to a neighbor's aid in such cases. But that does not mean that Jewish law regarded this duty to rescue as supererogatory or as conduct whose omission was not wrong. Failure to aid another in distress when one could do so at no risk to oneself was a sin before God. And within the *halakhic* system, this requirement was actively supported by rules exempting the rescuer from other civil and religious duties as well as from liability for harm inadvertently committed while trying to be of assistance. In this respect, Jewish religious law pioneered in the creation of a flexible legal structure able to support the moral injunction to beneficence ([15], pp. 218ff).

As strenuously as they insisted on the duty to help others in danger, however, the Rabbis set a limit on what was required. Precisely because of the nearly infinite value they placed on human life, they looked askance on any conduct – beneficent or otherwise – which exposed the agent to serious danger. In the Rabbis' eyes, preservation of one's own life was not just a counsel of prudence, but an urgent duty to oneself and to God. This duty was founded on the Biblical commandments "*Take heed to thyself and keep thy soul diligently*" (Deuteronomy 4:9) and "*Take ye therefore good heed unto yourselves*" (Deuteronomy 4: 15). In elaboration of this sensibility, an entire section of the Code of Laws is devoted to a detailed enumeration of actions which must be avoided because they endanger one's own life ([9], pp. 108ff). Similarly, the Rabbis insisted on a patient's cooperation with medical treatment of proven efficacy, and they regarded a failure to seek or to employ medical assistance as gravely sinful ([2], p. 28). They condemned suicide, euthanasia, or any hastening of death no matter how great the suffering continued life might represent,[7] and even self-destruction to avoid religious persecution and torture was prohibited unless it was clear that execution would also immediately follow ([8] and [17]).

These strong and repeated affirmations of the duty to protect and preserve one's own life had several understandable exceptions. As I have already noted, even if one was forced to do so at the threat of death, one could not legitimately kill another person. This teaching is justified in the Talmud by appeal to the sanctity and equality of life. To one who came before the teacher Raba and said to him, "The governor of my town has ordered me, 'Go and kill so and so; if not, I will slay thee,' Raba answered: "Let him rather slay you . . . ; who knows that your blood is redder? Perhaps his blood is redder" ([6], *Sanhedrin*, 74a). In fact, more than the equality of life is involved here, for a strict valuation of life can also validate self-protection in such cases. Rather, we see here what I take to be a very characteristic Rabbinic concern with precedent. The known fact that under duress it is considered legitimate to kill others is an open invitation to abuse, both on the part of those who would terrorize others at one remove and on the part of those who would use this reasoning to excuse their own homicidal acts.

A similar concern with precedent seems to underlie a related but even more striking departure from the Rabbinic insistence on self-protection. This involves the hypothetical case mentioned in the Talmud where desperadoes tell a group of Jewish travellers that unless they hand over one of their women for defilement, the whole group will be attacked. In such instances, the Rabbis maintain, it is incumbent upon all to die rather than to relinquish a single individual ([4], *Terumoth*, 8: 11–12). Superficially regarded, this appears to enjoin the most sacrificial beneficence, but more closely examined, I think, we really see here a concern with the long range or precedent implications of any alternative policy. Surely, the permission to hand over members of one's group under duress invites abuse by petty despots seeking to divide and conquer a community, and it also has terribly divisive implications for the conduct of group life as well. Against this, a policy of staunch resistance to such ultimata, no matter how foolish this resistance may sometimes seem, is one of the most powerful deterrents to their being issued in the first place. Thus, in this case and the one preceding we are dealing with only the most limited and understandable exceptions to the Rabbis' usual unwillingness to allow or to require an individual to risk his life on others' behalf.[8] As a general rule, the Rabbis were extremely reluctant to induce an individual to sacrifice himself for another person. When cases of conflict of life with life arose, as in the famous instance of the single pitcher of water, they appear to have allowed the vicissitudes of fate to determine the outcome.

So long as complex issues of precedent did not arise, they both allowed and encouraged an individual to look after himself.

It is important here, however, to indicate a minority tradition that does appear to enjoin sacrificial beneficence. A number of Jewish scholarly sources mention a view attributed to the Palestinian Talmud which requires a person to risk (but not clearly to sacrifice) his life in order to save the life of another person ([1], p. 384, n. 1). The great codifier of Jewish Law, R. Joseph Caro, supports this view, interpreting it to apply to cases where there is a *possibility* of serious harm or death to oneself but a certainty of harm or death to one's fellow. In such cases, he maintains, " . . . the one faces a certain danger, whereas the other [i.e., the rescuer] takes only doubtful risk, and whoever preserves the life of one person in Israel is as if he had preserved the entire world" ([13], p. 96). Nevertheless, despite the respect usually accorded Caro's views, because his position here is not given support in the more authoritative Babylonian Talmud, Caro's understanding is seen to be without solid foundation. As a result, the majority of influential writers do not appear to regard risking one's life on others' behalf as obligatory ([1], p. 384; [13], p. 97).

Is it, then, at least permissible to risk one's life for a fellow human being? Surprisingly, there is some doubt about this among Jewish authorities. The question was raised during the medieval period by a case and by a series of subsequent rulings which anticipate the issue of organ transplantation in our own day. R. Menahem Recanati, an Italian legalist and mystic of the late thirteenth and early fourteenth centuries started the debate with a ruling:

If a tyrant says to a Jew, 'Allow me to amputate one of your limbs' (an amputation which represents no danger to life), 'or else I will kill your fellow Jew,' some [authorities] say that he is obligated to allow his limb to be amputated since he would not die ([15], p. 211).

Two hundred and fifty years later, this decision is quoted disapprovingly by Rabbi David B. Zimra (*Radbaz*, d. 1573). He rejects the proof-text offered by R. Recanati and insists that no valid precedent can be found for this obligation. Expressing what seems to have become the normative Rabbinic view since his time, he affirms that in such cases the risk to one's own life rules out any obligation of this sort. His authoritative opinion concludes:

I, therefore, see no justification for his decision. It is an act of saintliness (*middat hasidût*) [i.e., above and beyond the legal requirement], and happy is the man who can live up to it. If, however, there is a possible risk of life, then [one who agrees to the amputation] is a foolish saint (*hasid shoteh*), for the *possible* danger to oneself takes precedence over the certain danger to one's fellow ([15], p. 212).

The latter part of this ruling graphically expresses Rabbinic doubts about even the *permissibility* of risky acts of beneficence. The sanctity of life permits extreme acts of self-sacrifice (e.g., loss of a limb) when these do not jeopardize the life of one who renders assistance. Such acts are not obligatory; they go beyond the requirements of law but are regarded as saintly. When, however, an act imperils the agent's life, the Rabbis hesitate. One who exposes his life to danger, even if there is only the possibility of death for him versus certainty of death for the other, is a 'foolish saint'. He is an individual at once misguided and excessive in his moral and religious devotion. Thus, elsewhere a 'foolish saint' is described as one who "brings destruction upon the world," and he is likened to the individual who looks upon a drowning woman and says, "It is improper for me to look upon her and rescue her" ([14], *Soṭah*, 3: 4; [6], *Soṭah*, 21b). Although this ruling by *Radbaz* has certain limitations – after all, it too deals with difficult matters of precedent – and while there are other points of view on this issue to be found in the literature, we can nevertheless conclude that dangerous heroic acts are largely discouraged by the mainstream of Jewish teaching. True, these acts are not clearly forbidden, and there are countervailing currents that insist that "everything depends on the circumstances" or that "one must not overly protect oneself" ([15], p. 211). But the dominant tendency is to look unfavorably on extreme beneficence of this sort. Despite the claims of Louis Jacobs, therefore, Sidney Carton and others like him who lay down their lives for a friend are not, for normative Judaism, to be regarded as saintly models.

III.

Before turning to some of the implications of this discussion for contemporary issues in medical ethics, I might try to summarize the broad structure of Jewish thinking I have developed on the matter of beneficence and self-sacrifice. We have seen that, in a very general way, Jewish law and teaching have been quite insistent about the duty of beneficence and the obligation to aid others in need. Indeed, the obligation to beneficence is not just left at the level of exhortation but is in some respects written into the structure of Jewish law so that infractions are brought under legal control and supporting norms are elaborated. In the area of financial or material assistance, we know, charitable giving at some cost to oneself is a concrete obligation of the religious life. To prevent profligacy, limits are placed on the obligation, but these very limits serve to point up the domain of clear requirement. Where another person's life is in peril, this religious obligation receives over-arching support from the Biblical commandment "*Thou shalt not stand idly by the blood of thy neighbor.*" Failure to help another in such distress, when one can do so at no serious risk to oneself, violates a positive commandment and is a grave sin. Thus, according to Jewish law, the onlookers in New York who some years ago failed to telephone the police from the safety of their apartments when the young woman, Kitty Genovese, was being physically assaulted, were guilty of a grave wrong. In this respect, Jewish religious law is more explicit on this matter than most civil law today.

At the far extreme, however, where assisting another risks one's own life, Jewish teaching becomes negative: one who exposes himself to serious risk (risk equivalent to amputation under the circumstances of medieval hygiene and technique) is held to be a 'foolish saint', one whose piety and morals are notable but are also excessive or overdone. In circumstances of lesser risk than this, we may suppose that the discouragement to beneficence wanes and, at some point, becomes encouragement. Beneficent conduct involving only modest risk to oneself but of lifesaving significance for the other is praised as saintly. Where no serious risk to oneself is involved and another's life is at issue, encouragement becomes injunction and a failure to aid is sinful. Finally, cutting across this whole structure are the specific duties of beneficence incumbent upon the physician (or equivalent health professional) by virtue of his role. At least for those patients clearly dependent upon his care, the physician is regarded as having a healing responsibility that may require him to undergo some risk on his patients' behalf. Thus, a physician is allowed to enter an infectious environment to succor a patient, and one who refuses to run this risk on behalf of his patients would be regarded as negligent in his duty.

IV.

Against the framework of this broad structure, we might consider the implications of Jewish teaching for some of the new issues in medical care that raise questions about the extent and limits of beneficence. In fact, some implications of traditional Jewish thinking for this area have already been developed by Jewish scholars, either in the classical or modern period, while other implications remain matters of conjecture and speculation. Among the more concrete and historically well-developed implications of the general stress on beneficence in Jewish law, for example, is the strong emphasis on one's duty to visit and comfort the sick. This duty is considered one of the most meritorious acts of beneficence, and, in a Talmudic passage recited daily by observant Jews, it is listed as one of the ten ethical or devotional exercises whose fruit a person enjoys in this life as well as the next ([6], *Shabbath*, 127a). One who visits the sick is likened to God ([6], *Soṭah*, 14a), and one who is negligent in this duty is held guilty of jeopardizing human life since visits help the sick regain their health ([6], *Nedarim*, 40a). As we might expect, however, risk to one's own life was regarded as a valid reason for not fulfilling this duty personally. Perhaps out of regard for both the sick individual and the visitor, the Talmud warns against calling on a bed-ridden person "before the fever has left him" ([6], *Nedarim*, 41a). Later Rabbinic teaching generally upheld the validity of self-protection in cases of epidemics ([13], p. 108). These exceptions aside, for the traditional Jew this duty has literal meaning and grounds a specific obligation of visitation. We might extrapolate from this to conjecture that this same duty can undergird a host of more contemporary obligations to the sick. Whatever one might do in a direct way to help the sick – ranging from rendering financial support to medical institutions to volunteering in programs that comfort patients or aiding by means of blood donation – might be seen as based upon this older obligation of visitation applied now to a modern medical setting.

While Jewish teaching has been somewhat explicit about beneficent conduct involving traditional matters, there is understandably far less discussion of issues related to some of the newer dimensions of medical practice. For example, there is relatively little Jewish treatment of the issue of participation in medical experimentation and only slightly more discussion of the matter of organ donation. In exploring these themes, it is even more necessary to extrapolate from the general structure of Jewish teaching on beneficence and self-sacrifice. Thus, where medical experimentation is concerned, traditional thinking would suggest the importance of a distinction, on the one hand, between therapeutic and non-therapeutic experimentation (that which aims at treating the subject for an illness he is suffering and that which aims at the expansion of medical knowledge generally, apart from the needs of the subject as patient) and, on the other hand, a distinction between experimentation that poses no serious risk to the experimental subject and that which places the subject's life or health in peril. This gives us a four-fold schema: (1) therapeutic experimentation involving no risk to the patient; (2) risky therapeutic experimentation; (3) relatively safe non-therapeutic experimentation; and (4) hazardous non-therapeutic experimentation.

Everything I have said about Jewish teaching concerning the duty to aid would suggest that participation in the first three of these forms of experimentation would be permissible for the devout Jew, and in some cases, participation might even be obligatory. A seriously ill person who fails to take medication of proven efficacy is regarded by Jewish teaching as violative of the duty to preserve his own life. Now, while this obligation was not usually taken to extend to medications or procedures of unproven value ([2], p. 28), a sick individual is certainly permitted to expose himself to untried means of therapy when these pose no further risk to his health or when they promise to restore him to health. Indeed, in the latter case, Jewish law would permit employment of risky and untried experiments that endangered the patient's life if it was felt that they might reverse the course of a fatal illness and restore him to health.[9]

Similarly, the Jewish stress on the duty of beneficence and the sanctity of life would appear to permit, and in some cases even require, participation in non-hazardous experimentation which aimed at improving medical knowledge generally. Rabbi Immanuel Jakobovitz supports this view when he maintains that if an experiment "involves no hazard to life or health, the obligation to volunteer for it devolves on anyone who may thereby help to promote the health interests of others" ([14], p. 382). In view of the emphasis in Jewish teaching on the duty to aid an actual person in distress (*a holeh lefaneinu* or literally, "one before us"), this responsibility would become even stronger if research was aimed at saving the life of an identifiable person.

From the Jewish point of view, therefore, it is only the fourth category, participation in hazardous experimental work not clearly of benefit to the research subject, that poses really serious questions. Everything that I have said about Jewish teaching to this point suggests that however noble such participation may appear to other ethical traditions, this kind of self-sacrificial beneficence is suspect from the Jewish perspective. Although there is dispute over this matter, the drift of Jewish teaching is to discourage such hazardous acts of beneficence. This discouragement might be less where an identifiable person's life could be saved by the research, but only marginally so.[10] Because of its very strong emphasis on the sanctity of life and one's duty to preserve one's own life, in other words, Judaism seems prepared to forego the benefits that research of this sort might provide. There is, however, one possible exception to this rule. A physician who is not ill himself might be permitted to act as subject in his own hazardous medical experiments if he believed he could thereby save the lives of future patients. This exception may derive from the special institutional responsibilities of the physician recognized by Judaism. Jewish teaching, therefore, may not prohibit and may even encourage the kind of risky self-experimentation undertaken by Walter Reed's medical colleagues.

In contrast to medical experimentation, the kind of beneficence represented by organ donation has received a good deal of explicit discussion in recent Jewish writings. This is probably because the related issues of amputation, self-mutilation, and the treatment of the deceased have long been subjects of Jewish legal thinking. Thus, a number of traditional norms bear on this area to forbid the mutilation of cadavers or living donors unless the prospect of saving an identifiable person or class of persons is involved.[11] Even where lives can be saved, however, Jewish teaching places limits on an individual's willingness to risk his life for another. Especially relevant here is the longstanding Jewish opposition to amputation of oneself on the neighbor's behalf. Recently, for example, the Sephardic Chief Rabbi of Israel, Ovadia Yosef, applied the classical position of Rabbi David B. Zimra to this matter. In doing so he found two governing considerations: (1) An individual is not obligated to put his own life in danger to save another, and (2) one may not donate a vital limb or organ if the transplantation represents a serious threat to one's own life. While these considerations may have reduced the permissibility of organ donation until recently, Rabbi Yosef reasoned, greater familiarity with the procedures and the vastly reduced risks of kidney donation today render donation of this organ permissible and even laudatory in most cases:

> ... according to information we have received from competent and God-fearing physicians, the danger [to the donor] involved in extracting a kidney is generally very small. Inasmuch as *Radbaz* and those of his school hold, therefore, that under such circumstances the *mizvāh*, *Thou shalt not stand idly by*, obtains, it follows that we must allow a healthy person to donate one of his kidneys, to save his fellow Israelite whose life is seriously threatened by a disease of the kidneys. Great is the *mizvāh* of saving human life, and it ... will afford the donor the protection of a thousand shields. In any event, the donation must be performed by competent physicians, and he who fulfills a *mizvāh* shall know no evil ([15], p. 212).

This reasoning gives scope for some further speculation on a Jewish estimate of organ donation. When the procedure poses serious or unknown risk to life (as kidney donation may have in its very beginnings where the donor was concerned), traditional Jewish thinking would tend to be prohibitory.[12] Where a procedure is without risk to the donor but vital to the recipient, as blood or marrow donation may be in our own day, Jewish teaching would appear to require assistance by whoever is in a position to help. And where the hazards are intermediate between these extremes, Judaism both encourages and praises the beneficent conduct involved.

V.

We can see, therefore, that the Jewish religious-ethical tradition possesses a remarkably well-developed structure of normative teaching with regard to beneficence in general and medical beneficence in particular. Not only is this structure complex, it is to some degree distinctive when set against other religious positions, especially that of Christianity. Like Christianity, for example, Judaism strongly encourages many kinds of beneficent acts, but because of the essentially legal and juridical nature of Jewish religious teaching – Jewish religious law, after all, is meant to be the foundation for a theocratic political order – these encouragements are made concrete through a series of specific normative enactments. Where risk-free vital assistance is involved, therefore, Judaism goes beyond exhortation and legally requires and facilitates assistance to the other. On the opposite side of the spectrum, where assistance can only be rendered at substantial risk to the self, the distinction between Judaism and Christianity is even sharper. Although St. Thomas Aquinas could advocate self-sacrifice on others' behalf for the sake of the 'virtue' involved in the act (*Summa Theologica*, II–II, Q. 26, Art. 5), the major Jewish view withholds social support or encouragement for such conduct.

In might be thought that this structure of Jewish norms is overly constraining. Surely the broader Christian position with its strong positive advocacy of beneficence and its greater reliance on individual decision at each point along the spectrum has appeal and value. It places significant emphasis upon the individual's own freedom to decide. But even as this is acknowledged, it must be kept in mind that the Jewish position has value as well and in its own way is supportive of human freedom. For one thing, part of the great significance of the entire *halakhic* tradition is the effort to embody lofty moral ideals in a structure of laws and social institutions needed to give them life. Jewish law aimed to make moral vision moral reality, and in this case its teachings helped transform benevolent impulses into a pattern of beneficent acts.

Freedom, too, was served by this structure. At first sight, this may seem paradoxical. How can an individual's freedom to determine his own conduct be facilitated by norms prohibiting, for example, extreme acts of beneficence? Does it aid freedom to prevent an individual from laying down his life on another's behalf? Why not instead leave this matter entirely to the individual's own determination? In responding to these questions, and in trying to see the sense of Jewish teaching on this matter, it is useful to keep in mind the fact that freedom can sometimes itself be a constraint. The history of kidney transplantation reveals that individuals in a position to donate one of their kidneys have sometimes felt themselves subject to conflicting pressures. On the one hand, they wish to render lifesaving aid to a relative. On the other hand, they find themselves naturally fearful for their own well-being or they feel a responsibility to their own families who are dependent on them. In this situation whichever choice the individual makes, he can be beset with fear, anxiety, or guilt.[13] In a moral context where expectations are high, what is supererogatory can quickly become a duty.

These concerns, of course, are the very ones that have motivated Jewish thinking about beneficence. Recognizing the enormous value that each person's life represents – both to himself and to his loved ones – Judaism refrains from pressing beneficence to the limit. Instead, it chooses to fabricate a protective structure of norms that create an island of freedom from coercion. This freedom is also brought into being by the tradition's willingness to voice a clear "no" to forms of conduct into which excessive zeal or social pressure can force an individual against his own deepest wishes. Indeed, much the same approach is used by those Jewish teachers who actively prohibit religious martyrdom in all but the most unavoidable cases.[14] In matters as subtle as these, of course, no one can say whether this approach is superior or inferior to its alternatives. But we can say that Jewish thinking on this problem represents an important and distinctive tradition of moral reflection. It is a tradition, moreover, that merits continuing attention as we seek to come to terms with the new opportunities for beneficence opened up by modern medical practice.

Dartmouth College
Hanover, New Hampshire

NOTES

[1] All translations are drawn from [6].
[2] Ha'am's position is shared by Petuchowski [16] and by J. H. Hertz in [10], pp. 564ff.
[3] See his article in [12], pp. 41–47 and his book [11], pp. 34ff.
[4] See, for example, [6], *Baba Bathra*, 8a–11a.
[5] Some sources omit the words "from Israel" from this passage.
[6] This obligation was also grounded in the scriptural exhortation "And thou shalt return it to him" (Deuteronomy 22: 2), which, on the basis of a pleonasm in the Hebrew text, the Rabbis interpreted to apply to the restoration of the neighbor's body as well as his property.
[7] Although the Rabbis did not permit hastening of death, they allowed the cessation of

therapy for a *goses* or one already in the dying process. See [3], *Yoreh De' ah* 339: 1 and [2], p. 33.

8 The fact that a concern with precedent is operative here is suggested by the Rabbis' willingness to permit an individual to be handed over – whether objectively guilty or not – if this specific person was called for. Presumably, the deterrent effect of mass resistence would be substantially lessened in this case. For a further discussion of this matter, see [5].

9 This teaching was extrapolated from analogous cases of non-medical rescue mentioned in the Talmud, especially in [6], *Yoma*, 85a and *'Abodah Zarah*, 27b.

10 This is the view of Jakobovits who states that "Possibly hazardous experiments may be performed on humans only if they may be potentially helpful to the subject himself . . . ," ([14], p. 382). Drawing more heavily on the somewhat less authoritative position of Caro, Bleich disagrees with Jakobovits on this matter and maintains that participation in hazardous, non-therapeutic experimentation is permissible for the Jew and perhaps even praiseworthy ([1], p. 385).

11 These norms are based on Talmudic prohibitions on desecration of the dead ([6], *Sanhedrin*, 47a and *Hullin*, 11b). See [19], pp. 331–348 and [13], pp. 97, 126–152.

12 For a different view see [18], p. 366.

13 Although their findings support the view that kidney donation can be a very meaningful decision for those called upon to help a relative, Carl H. Fellner and John R. Marshall's study ([7], pp. 269–281) also points up the considerable anxiety experienced by some who are called upon in these circumstances.

14 See [13], p. 53.

BIBLIOGRAPHY

1. Bleich, J. D.: 1979, 'Experimentation on Human Subjects', in F. Rosner and J. D. Bleich (eds.), *Jewish Bioethics*, Sanhedrin Press, New York, pp. 384–386.
2. Bleich, J. D.: 1979, 'The Obligation to Heal in the Judaic Tradition: A Comparative Analysis', in F. Rosner and J. D. Bleich (eds.), *Jewish Bioethics*, Sanhedrin Press, New York, pp. 1–44.
3. Caro, J.: 1954, *Yoreh De' ah, Code of Hebrew Law*, C. N. Denburg (transl.), Jurisprudence Press, Montreal.
4. Danby, H., (trans.): 1933, *The Mishnah*, Oxford University Press, Oxford.
5. Daube, D.: 1965, *Collaboration With Tyranny in Jewish Law*, Oxford University Press, London.
6. Epstein, I. (ed.): 1935–1961, *Babylonian Talmud*, Soncino Press, London.
7. Fellner, C. H. and Marshall, J. R.: 1970, 'Kidney Donors', in J. Macaulay and L. Berkowitz (eds.), *Altruism and Helping Behavior*, Academic Press, New York, pp. 269–281.
8. Frimer, D. I.: 1975, 'Masada – in the Light of Halakhah', *Tradition* **12**, 27–43.
9. Goldin, H. (transl.): 1961, *Code of Jewish Law*, Hebrew Publ. Co., New York.
10. Hertz, J. H.: 1941, *The Pentateuch and Haftorahs*, Vol. I, Metzudah Publ., London.
11. Jacobs, L.: 1964, *Principles of Jewish Faith*, KTAV, New York.

12. Jacobs, L.: 1957, 'Greater Love Hath No Man . . . The Jewish Point of View on Self-Sacrifice', *Judaism* **6**, 41–47.
13. Jakobovits, I.: 1959, *Jewish Medical Ethics*, Bloch Publ. Co., New York.
14. Jakobovits, I.: 1979, 'Medical Experimentation on Humans in Jewish Law', in F. Rosen and J. D. Bleich (eds.), *Jewish Bioethics*, Sanhedrin Press, New York, pp. 377–383.
15. Kirschenbaum, A.: 1980, 'The Bystander's Duty to Rescue in Jewish Law', *Journal of Religious Ethics* **8**, 204–226.
16. Petuchowski, J.: 1955, 'The Limits of Self-Sacrifice', in M. Fox (ed.), *Modern Jewish Ethics*, Ohio State University Press, pp. 103–118.
17. Rosner, F.: 1979, 'Suicide in Jewish Law', in F. Rosner and J. D. Bleich (eds.), *Jewish Bioethics*, Sanhedrin Press, New York, pp. 317–330.
18. Rosner, F.: 1979, 'Organ Transplantation in Jewish Law', in F. Rosner and J. D. Bleich (eds.), *Jewish Bioethics*, Sanhedrin Press, New York, pp. 358–374.
19. Rosner, F.: 1979, 'Autopsy in Jewish Law and the Israeli Autopsy Controversy', in F. Rosner and J. D. Bleich (eds.), *Jewish Bioethics*, Sanhedrin Press, New York, pp. 331–348.

WILLIAM E. MAY

ROMAN CATHOLIC ETHICS AND BENEFICENCE

I. INTRODUCTION

There is ambiguity in both 'Roman Catholic ethics' and 'beneficence'. The former can mean either the body of moral norms, both general and specific, authoritatively proposed by those who hold the teaching office or magisterium within the Roman Catholic Church, namely the pope and the bishops throughout the world in communion with him,[1] or the diverse ethical theories, along with their substantive conclusions, that have been developed by various Roman Catholic moral theologians and philosophers. This paper will attempt to describe both the moral teachings authoritatively proposed by the magisterium, particularly with reference to issues relating to health care, and the debate currently taking place among Roman Catholic writers over normative ethical principles and the sorts of human choices and actions justifiable in terms of these principles, particularly with reference to health care questions.

'Beneficence', too, has its ambiguities. In traditional Roman Catholic thought, this term has not been used to describe a principle of morality, as it is by many contemporary moral philosophers. Rather the term has been used by Roman Catholic writers to designate an act of virtue, principally an act of the virtue of charity or love, of effectively doing the good we will to another if the capacity to do this good lies within us ([1], 2–2, q. 31, a. 1).

When beneficence is used by Roman Catholic writers to describe an act of the virtue of charity, one must not conclude that acting in this way is merely something supererogatory and not obligatory. For to have good will to all, and effectively to do good to them, is regarded by Roman Catholic writers as a commandment from God himself. Even enemies are to be loved, and good is to be done to them, for as St. Augustine said, "each one of us ought so to treat a man as we wish him to be, even if he has not yet become what we wish" (*Contra Mendacium*, c. 6. 15, trans. Jaffe, in [2], p. 141), and we are to wish him to be a friend of God and, in God, a friend of ours.

Moreover, if the 'good' we are to do to others is a 'good' owed to them because they have a right to it, beneficence, as an act of virtue, is required not

Earl E. Shelp (ed.), Beneficence and Health Care, 127–151.

only by reason of charity but also by reason of justice. It is in this sense that we are morally required, in traditional Roman Catholic moral thought, to do good, in the sense of establishing justice between ourselves and others, and to refrain from doing evil, in the sense of preserving a just order that has already been established ([1], 2–2, q. 79, a. 1).

Historically, beneficence as an act of the virtue of charity is of central significance for health care. The historian, E. Nasalli-Rocca, has observed that "the history of hospitals has been shaped by principles in accord with the teachings of Christ and the commandment of fraternal charity" ([59], p. 160). In response to the requirements of charity the Catholic Church early in its history established *xenodochia* or inns for travelers and *nosocomia* or infirmaries, which were the forerunners of modern hospitals and clinics, and it is in response to these requirements that the Catholic Church today maintains health-care facilities, whose primary concern, the bishops of the United States affirmed in 1971, is to serve the "total good of the patient". So central is this beneficent concern for Catholic health-care facilities that, should any Catholic institution find itself incapable of fulfilling this basic mission, the institution in question "would have no justification for continuing its existence as a Catholic health facility" ([12], p. 206).

In current philosophical moral literature, 'beneficence' is taken in various senses. One way in which it is understood is as a principle prescribing that we are to do the good and to avoid or refrain from doing harm ([21], p. 45), and so understood 'beneficence' is simply another way of describing what the Roman Catholic tradition has called the first principle or precept of natural law.[2] But 'beneficence' is also used in current literature to include not only the prescription to do good and avoid evil but also the prescription to bring about the best possible balance of good over evil ([3], p. 136). Understanding 'beneficence' in this way, some modern philosophers then argue that we must distinguish two principles under the general rubric of beneficence, the first requiring the provision of benefits and the second the balancing of benefits and harms. The second of these principles is commonly called the principle of utility. When 'beneficence' is used in this way, it is, as we shall see, a notion that seems to be congenial to the thought of a number of contemporary Roman Catholic writers, who nonetheless eschew the language of utility and much prefer to speak about the principle of proportionality or of proportionate good or reason. Other contemporary Roman Catholic writers vigorously reject both the principle of utility and the principle of proportionate good. The latter, while firmly committed to 'beneficence' as one way of describing the first principle of practical reason or of natural law, regard 'beneficence' taken in this second sense as a pseudo-principle, one in their judgment incompatible with a correct understanding of the principle that good is to be done and pursued and evil is to be avoided and one that, in their view, leads to conclusions directly contrary to the firm teaching of the magisterium of the Church on moral questions.

Finally, there is considerable debate, both among moral philosophers at large and among Roman Catholic writers, whether one can properly use the term 'beneficence' to describe certain sorts of human choices and actions, many of central importance to health care. For instance, is it accurate to speak of 'beneficent' euthanasia or infanticide? Can the saline-induced abortion of an eighteen-week-old fetus, even one that may have been diagnosed to be suffering from a gravely impairing genetic malady, rightly be described as an act of beneficence? Such queries could be indefinitely extended.

My purpose now is the following: first, to examine some relevant teachings of the magisterium of the Roman Catholic Church on moral issues, noting in particular some specific teachings pertinent to health care; second, to provide a critical account of the debate among Roman Catholic writers concerning the 'principle of proportionality'; and third, to conclude with personal reflections on Roman Catholic ethics, beneficence, and health care.

II. ROMAN CATHOLIC MORAL TEACHING

The magisterium of the Roman Catholic Church does not propose a particular moral theory, because it is not its function to serve as a philosophical institute. Nor are those who exercise the magisterium necessarily moral philosophers or theologians. Yet it does, in carrying out its proper mission of proclaiming the gospel of Christ, make some claims about what human beings are to do if they are to be the beings they are meant to be. These claims the magisterium holds to be true, and they include both general and specific propositions. My purpose is not to offer any arguments in support of these claims; it is rather to describe them and to note some of their more important features.

First, let us take some claims about issues usually regarded as metaethical in contemporary literature. Here the more significant claims seem to be the following:

(1) There is an objective moral order according to which human actions are to be judged as good or evil.[3]

(2) This objective moral order has God as its ultimate source, and his "divine law, eternal, objective, and universal" is the "highest norm of human life" ([77], n. 3, p. 801).

(3) Humanity has been so constituted that one is impelled by one's nature and bound by a moral obligation to seek the truth about what one is to do and to direct one's entire life in accord with the demands of truth ([77], n. 2, pp. 800–801).

(4) Humanity is therefore capable of coming to know the highest norm of human life, namely God's 'law,' and one's mode of consciously participating in it is properly called 'natural law' ([77], n. 2, pp. 800–801; cf. also, [78], n. 16).

(5) This 'law', natural and divine, is not limited to one principle but rather contains many principles regulative of human choices and actions, principles that are universally binding on individuals and human communities ([78], n. 79, p. 988).[4]

These propositions show clearly that Roman Catholic teaching on ethical questions is uncompromisingly cognitivist and unalterably opposed to relativism. This Church unwaveringly teaches that the ultimate dignity of the human person consists in the human capacity (a) to discover the objective truth about what is to be done, both in particular judgments about what one is to do here and now and with respect to general norms ([78], n. 16, p. 916), and (b) to choose freely to "do the truth," that is, to do the good that one discovers one is obliged to do and to refrain from doing evil ([78], n. 17, p. 916).[5]

Note that one of the propositions asserted as true by the magisterium of the Church affirms that there are several principles, universally binding, of the 'natural law'. Since the magisterium of the Church does not pretend to do the work proper to moral philosophers and theologians (its mission is to proclaim the saving truths committed to it by Christ the Lord),[6] it would be foolish to expect to find in its teachings any effort rigorously to articulate a taxonomic list of such principles. Yet one does find in its teachings appeals to universal normative principles and goods of human existence. Thus the magisterium teaches that good is to be done and that evil is to be avoided.[7] It insists that there are real goods of human persons, *goods that ought to be valued* by all humanity, not goods that are valuable only because they are, in fact, actually deemed valuable by some. Among such goods are those of human life itself, justice, truth, love and peace, goods that are truly "fruits of our nature," and goods that we will find, transfigured, in God's kingdom of "truth and life, holiness and grace" ([78], n. 39, p. 938).[8]

In addition, the magisterium of the Roman Catholic Church clearly affirms, in company with St. Paul (cf. Romans 3: 8) that we are not to do evil so that good may come about ([62], p. 45 and [61], pp. 337–338).

By this it means that we are not, of deliberate purpose, to choose to destroy, damage, or impede anything that is truly a good of a human person in order that we might, thereby, be able to participate in or protect something else that is good. The magisterium thus seems to indicate that one of the universally binding and unchanging principles of natural law is that evil is not to be done so that good may come about.

It is in terms of these general claims that the magisterium of the Roman Catholic Church proposes more specific norms regulative of human choices and actions. Thus, while recognizing that a nation has a right to defend itself legitimately against attack and that military personnel can properly regard themselves as "custodians of the security and freedom of their fellow-countrymen," the Church firmly teaches that certain actions in war violate "the natural law of peoples and its universal principles." Specifically, it teaches that "*every* act of war directed to the indiscriminate destruction of whole cities or vast areas with their inhabitants is a crime against God and man, which merits firm and unequivocal condemnation" ([78], n. 80, p. 990). Through this specific norm the magisterium absolutely condemns as morally wicked, and as totally opposed to principles that must be observed in waging war justly, any choice to defend a nation's security and resist the enemy by direct attack upon noncombatants.

Again, in speaking of the inherent dignity of human persons and the respect due to human persons, the subjects in whom the goods previously mentioned are meant to flourish, the magisterium of the Roman Catholic Church firmly teaches that

> all offenses against life itself, such as murder, genocide, abortion, euthanasia and willful suicide; all violations of the integrity of the human person, such as mutilation, physical and mental torture, undue psychological pressures; all offenses against human dignity, such as subhuman living conditions, arbitrary imprisonment, deportation, slavery, prostitution, the selling of women and children, degrading working conditions where men are treated as mere tools for profit rather than free and responsible persons; *all* of these and the like are criminal: they poison civilization; and they debase the perpetrators more than the victims and militate against the honor of the creator ([78], n. 27, p. 928).

Evidently the magisterium of the Church believes that the sorts of human acts described here are completely opposed to the universally binding principles of the 'natural law' and to the 'highest norm of human life', namely God's divine law, eternal, objective, and universal. It should be noted that although the passage just cited at times uses morally evaluative terms in describing the sorts of acts that are morally criminal in nature (e.g., *murder, undue* psychological pressures, *subhuman* living conditions, *arbitrary* imprisonment), it likewise condemns absolutely other acts described in nonmoral terms (e.g., *euthanasia, genocide, abortion, suicide*). It is evident, too, that several of the sorts of actions absolutely condemned in this passage have immediate relevance to issues involved in health care, e.g., abortion, euthanasia, mutilation. The magisterium of the Roman Catholic Church clearly holds that certain sorts of choices – specifically the choices to abort unborn children, to kill patients mercifully (whether dying or nondying), to mutilate patients – are not capable of being reconciled with the requirements of the universally binding principles of the divine and natural law.[9] Thus in its judgment such choices and actions cannot count as 'beneficent' behavior.

In addition, the Roman Catholic magisterium unequivocally teaches that contraception and contraceptive sterilization also violate objective norms of morality. These actions, too, it holds, cannot rightfully be chosen as legitimate ways of avoiding pregnancies where there are serious reasons for preventing pregnancies from occurring.[10] Thus the magisterium clearly judges that these acts cannot rightly be regarded as 'beneficent' or as morally good means for fostering health care.

In developing its teaching on specific sorts of acts quite relevant to health care, the magisterium has made use of distinctions and subsidiary principles in order to clarify its teaching on the absolute immorality of specifiable sorts of acts. One key distinction that the magisterium has employed, when speaking about acts foreseen to cause evil, is that between what is directly intended and what is not so intended. Thus Pope Pius XII, in articulating the magisterium's firm teaching on the absolute immorality of abortion, an act judged to be utterly irreconcilable with the universally binding principles of the divine and natural law, spoke as follows:

> We have on purpose always used the expression "*direct* attack on the life of the innocent," "*direct* killing." For if, for instance, the safety of the life of the mother-to-be, independently of her pregnant condition, should urgently require a surgical operation or other therapeutic treatment, which would have as a side effect, *in no way willed or intended yet inevitable*, the death of the fetus, then such an act could not any longer be called a *direct* attack on innocent life. With these conditions, the operation, like other similar medical interventions, can be allowable, *always assuming that a good of great worth, such as life, is at stake, and that it is not possible to delay until after the baby is born or to make use* of some other effective remedy ([64], pp. 191–192).[11]

The distinction between what is directly intended and done and what is not directly intended and done, along with the other "assumptions" noted by Pius XII, are critical elements of the norm that has come to be called the 'principle of double effect'. Since the debate among Roman Catholic writers over the 'principle of proportionality' was stimulated by considerations of the 'principle of double effect', more detailed consideration of the 'direct-indirect' distinction and of the other "assumptions" noted by Pius XII will be given later, in the section devoted to contemporary debates among Roman Catholic writers. But it is necessary here to note that the magisterium clearly regards the distinction between what is 'directly' intended and what is not so intended as one of crucial moral, and not merely descriptive, importance. It unambiguously teaches that it is always morally wicked to choose to kill innocent human beings with direct intent. Acts executing such a choice are truly described morally as acts of killing innocent human life, and such acts – and similarly other acts in which one directly or precisely intends the evil – are always morally wrong because they violate the unchanging principles of the natural and divine law.[12] On the other hand, the magisterium teaches that an act that may in fact cause the death of an innocent human being (or some other evil to a human person) is not to be described morally as a human act of killing an innocent person (or of blinding, crippling, or otherwise maiming a human person) if the evil the act causes is *not* directly intended, *and* if the good the act does is what is directly intended, *and* if other conditions are met. It is rather to be described morally as a different sort of human act, that is, the execution of a different sort of moral choice. Thus, in the example considered by Pope Pius XII, the human act in question is *not* one of aborting but is rather one of protecting the life of the mother-to-be.

Again, particularly in teaching on the morality of medical procedures (e.g., surgery, radiation, chemotherapy, etc.) that 'mutilate' the patient, the magisterium has invoked the 'principle of totality'.[13] According to this principle, obviously one that the magisterium considers to be in conformity with the universally binding principles of divine and natural law, a person may rightfully choose to accept medical treatment, and doctors may rightfully choose to provide such treatment, that destroys or impedes the functioning of a bodily member or organ when the bodily part in question is diseased or when its continued functioning poses a serious threat to the life or health of the human person.[14] Thus this principle justifies various kinds of 'mutilating' acts. Still the magisterium has made it clear that, in a 'mutilation' such as sterilization, the principle of totality needs to be understood in conjunction with the 'direct-indirect' distinction. 'Direct' or 'contraceptive' sterilization, that is, a sterilization whose precise purpose, one necessarily intended directly or precisely, is to sterilize the patient and thereby make the patient incapable of generating human life, is not morally justifiable, nor can the 'principle of totality' be used as an alleged ground of justification. It cannot properly be used because in 'direct' sterilization, the evil, i.e., the deprivation of a human person's generative capacity, is directly intended. On the other hand, an act that results in the patient's sterilization is morally justifiable by the principle of totality if the purpose of the sterilizing procedure is to remove a pathological generative organ or to impede the functioning of a generative organ when its continued functioning poses a serious threat to the life or health of the patient, for instance, when the production of hormones contributes to the spread of cancer. In such instances the 'sterilization' is not what is directly intended. The act chosen renders the patient sterile, but it is not chosen precisely '*to* sterilize' the patient ([66], p. 673).[15]

The import of the 'direct-indirect' distinction in understanding how the magisterium employs the 'principle of totality' in questions concerning sterilization helps us to understand how this principle is used by the magisterium in other 'mutilating' situations. When a doctor amputates a cancerous leg, for instance, he or she is not setting out (directly intending) to make the patient one-legged, nor in excising a cancerous eye is he or she directly intending to make the patient blind, although in the first case the patient becomes one-legged and in the second he or she becomes blind. Still the 'mutilation' of the patient is not precisely or directly what is intended. Mutilation, as such, is among those human acts that Vatican Council II enumerates as being contrary to the integrity of the human person and hence as a violation of the universally binding principles of natural and divine law ([78], n. 27, p. 928).

This section has sought to describe the principal moral teachings of the Roman Catholic magisterium pertinent to any consideration of 'beneficence' and health care. The investigation will now center on the more significant developments in moral inquiry currently taking place within the Roman Catholic scholarly community germane to this inquiry.

III. CONTEMPORARY DEVELOPMENTS IN ROMAN CATHOLIC MORAL THOUGHT: PROPORTIONALISM AND BENEFICENCE

In current moral philosophy the 'principle of beneficence' is frequently associated with the 'principle of utility'. According to the latter we ought to "do the act or follow the practice or rule that will or probably will bring about the *greatest possible balance of good over evil* in the universe". As

Frankena notes, "the principle of utility represents a compromise with the ideal. The ideal is to do only good and not to do any harm . . . But this is often impossible, and then we seem forced to try to bring about the best possible balance of good over evil" ([21], p. 45).[16]

Remarkably similar language is used by many Roman Catholic moral theologians today. Thus Timothy E. O'Connell, in a work intended as a text in Catholic seminaries and institutions of higher learning and one praised by several colleagues as a synthesis of the more important developments in contemporary Roman Catholic moral theology,[17] writes as follows:

> The traditional moral maxim, do good and avoid evil, may have some colloquial significance. But as a precise description of the living of our lives it is simply and inevitably impossible . . . Our best hope is to do as much good as possible and as little evil as necessary ([60], p. 152).

O'Connell – and here he represents a distinct current of contemporary Roman Catholic moral thought that has become known as 'proportionalism'[18] – claims that a sound natural law approach to the subject of objective norms must admit that the natural law is real, experiential, consequential, historical, and proportional. By this he means that ultimately "specific actions are to be evaluated from a moral point of view by considering their actual effects, or consequences". While opposing what he terms a "microconsequentialism", he proposes a "macroconsequentialism", where one attempts to take into account the results of particular acts in *both* the near and long term. In developing his view that the true norm of morality must be not only consequentialistic but proportional, O'Connell claims that we discover the right thing to do "by balancing the various 'goods' and 'bads' that are part of the situation and by trying to achieve the greatest proportion of goods to bads". The right action is the one "which contains the proportionally greatest maximization of good and minimization of evil" ([60], pp. 147, 148, 153).

Roman Catholic writers who defend proportionalism differ on specific issues, but agree in holding that there are no sorts or kinds of actions, describable in morally neutral terms, that one ought never, under any circumstances and for any reason, choose to do. According to them anyone may truthfully judge that acts of any kind (including the act of deliberately killing an innocent human person with direct intent) are in some circumstances permissible and even obligatory in order to preserve and foster greater goods or to avoid greater evils, and anyone may rightfully choose to do these deeds.[19] They thus hold that there are no specific (concrete) moral norms proscribing certain specific sorts of acts that are absolutely unexceptional. They admit that there may be some specific negative norms that are "virtually exceptionless" ([34], pp. 50–51 and [55], pp. 433–434), or "practical guidelines" ([37], p. 97), but they deny that there are any universally true negative moral norms ([37], p. 97 and [55], pp. 576–591, 638–651, 747–750).

Catholic writers holding this position recognize that it is in principle capable of justifying certain sorts of human choices and acts that are judged by the magisterium to violate the universally binding principles of the natural and divine law, for example, directly intended abortion, mercy killing, contraception and contraceptive sterilization. They claim, however, that no specific moral teachings of the magisterium are infallibly proposed ([34], p. 54 and [40]), an aspect of their thought that is not of concern to our inquiry,[20] and that therefore their disagreement with or dissent from these teachings is legitimate and can be accepted as probably true opinions by Catholics.[21]

On the view of these writers, several sorts of procedures related to health care may be judged 'beneficent' although the Roman Catholic magisterium judges them to be quite otherwise. All of the Roman Catholic writers subscribing to proportionalism agree that there are, in fact, proportionate reasons why married couples should practice contraception and why some married couples may rightfully choose contraceptive sterilization in order to avoid pregnancies that may seriously threaten the health and life of the mother or that may result in the birth of a child severely crippled by a genetically induced malady.[22] Others argue that abortions may be directly intended and chosen in appropriate conditions,[23] and still others believe that mercy killing is truly beneficent in some situations,[24] and that any of these deeds is in principle capable of being morally justified. The key issue is whether there is or is not a 'proportionate' reason for choosing the act in question.[25]

The rise of the theory of 'proportionalism' within Roman Catholic moral thought has been stimulated by attempts on the part of Roman Catholic writers to reflect on the 'principle of double effect' and on the distinction between the directly intended and the indirectly intended, a principle and a distinction utilized by the Roman Catholic magisterium. It will be useful, therefore, to examine this principle and the debates it has engendered within Roman Catholicism.

As commonly formulated by older Roman Catholic writers, the 'principle of double effect' is a 'principle' that is to be used in helping one to determine whether or not one can rightfully choose to do a deed that causes both good and evil. According to this principle, one may rightfully cause evil through an act of choice if and only if four conditions are verified. These are: (1) the act itself, prescinding from the evil caused, is morally good or at least indifferent; (2) the good effect is what the agent intends directly, only permitting or 'indirectly' intending the evil effect; (3) the good effect must not come about by means of the evil effect; and (4) there must be some proportionately grave reason for permitting the evil effect to occur.[26]

The first of these conditions has not occasioned great controversy, but the others have. The second presupposes that there is a morally significant difference between an intending will and a permitting will. The third rejects the idea that a good end can justify an evil means, and in explaining this condition older Catholic writers held that it was grounded in divine revelation, particularly in St. Paul's teaching in Romans 3: 8.[27] The fourth uses the language of proportionality, and presupposes that there must be some way of determining whether or not there really is a 'proportionately grave reason' for permitting evil to occur.

In the contemporary discussion among Roman Catholic writers concerning this principle,[28] many have come to conclude that the fourth condition is what is truly morally decisive in questions with respect to moral choices that cause evil. Those who reach this conclusion thus hold that 'proportionality' or 'proportionate reason' more accurately shows what is morally at stake in the 'principle of double effect'. Some of these writers, for instance William Van der Marck[29] and Cornelius Van der Poel,[30] reject the meaningfulness of the distinction between the directly and the indirectly intended; others, including Richard A. McCormick ([51], pp. 64–72),[31] believe that this distinction, although not ultimately of crucial *moral* significance, has a *descriptive* value insofar as it indicates the way in which the human person, through the will, is related to the evil he or she freely chooses and does.

All of these writers hold that it is necessary to distinguish between 'moral' evil and what is variously called 'pre-moral', 'non-moral', or 'ontic' evil. 'Pre-moral' ('non-moral', 'ontic') evil is simply the deprivation of a good that is due to an entity. Thus death is the deprivation of the good of life, sickness of the good of health, blindness of the good of sight. 'Moral' evil, on the other hand, designates a morally evil human choice and act, theologically understood as the personal sin of the human agent. These writers hold that it is always *morally* wicked to choose to do a 'moral' evil, and thus that it is always morally wicked to intend that another human person sin or choose wickedly. Yet they argue that the choice to do 'ontic' evil (e.g., the choice to kill a human person, to intend that this person *be* dead) is morally wicked only when there is no proportionate reason to justify such choice. According to them, this choice can be morally *good* if a 'proportionate' reason exists for making this choice and intending this evil.[32] These writers thus admit that some sorts of choices and acts are 'intrinsically evil' in a weak sense, inasmuch as some sorts of human choices and acts necessarily include the intent to do 'ontic' evil, but they deny that there are any 'intrinsically evil' sorts of human choices and acts in the strong or *moral* sense.[33] Thus for these writers the proposition, 'the end does not justify the means' – a proposition implicit in the third condition of the 'principle of double effect' – must be rejected if by 'end' one means the consequences of one's act, for it is these consequences precisely that justify the means ([60], p. 172). They teach that one may rightfully intend evil ('ontic' evil) for the sake of good to come, or, as McCormick puts it, one may intend evil *in ordine ad finem proportionatum* ([55], pp. 355–356). These writers hold that the teaching of St. Paul in Romans 3: 8 must be understood as asserting that we are not to choose *moral* evil for the sake of good to come but that we are permitted and at times obliged to choose to do *ontic* evil for the sake of a greater good to come, even if the 'greater good' is a 'lesser evil'.[34]

Other contemporary Catholic writers (in company with other religious and secular authors)[35] vigorously oppose this development within Roman Catholic moral thought – even when they agree that older Catholic writers interpreted the third condition of the 'principle of double effect' too physicalistically and restrictively.[36] They offer several criticisms of 'proportionalism' or 'consequentialism'.

Of these criticisms the most important is that 'proportionalism' requires something impossible, namely the commensurating of the incommensurable and the calculating of the incalculable. The proportionalists argue that we are to choose that act which will result in the 'greater good'. This proposal has an initial plausibility, for it would obviously be absurd to choose the 'lesser good' or 'greater evil'.[37] Yet the plausibility of this proposal rests on an ambiguity in the term 'good'. The morally upright person surely is seeking to do the good, in the sense of the morally good or right action. Yet the proportionalist holds that it is possible to determine, prior to choice, which among diverse alternative choices is *morally* good by balancing or weighing or measuring the *nonmoral* (ontic) goods and evils that one's freely chosen act will cause. The difficulty here, as many writers have noted,[38] is that there is no unambiguous or univocal measure according to which the various goods in question (goods such as human life itself, truth, friendship, justice) can be compared. These goods of human persons, unlike useful goods such as money, clothes, food, etc., are simply incalculable and incommensurable. Although none is an absolute good, in the sense of the highest good or *summum bonum*, each is truly a good of human persons and of human flourishing, and as such a good to be prized and not priced, a good participating in the goodness of the human person and a good that functions, when known, as an intelligent principle of purposeful human activity and choice.[39] Yet, these authors argue, the proposal entertained by those Roman Catholic writers who advocate 'proportionalism' requires that one be able in some unambiguous and non-arbitrary way to judge, prior to choice, that some of these goods are 'greater' than others. But this proposal requires the impossible and is, therefore, an unreasonable and false proposal.[40]

Two different kinds of response have been given to this criticism. One advocate of proportionalism, Richard A. McCormick ([51], pp. 95–106), who previously asserted that the judgment of a proportionate reason required the commensurating of human goods and the establishment of them into a hierarchy, now admits that it is impossible to compare or measure the goods in question. Yet he insists that "while the basic goods are not commensurable (one *against* the other), they are clearly associated" or interrelated. Moreover, by considering these goods in their interrelationship one can make a judgment that the destruction or impeding of one good in the present circumstances will not lead to an undermining of that good and that its destruction or impediment here and now is *necessary* in order to foster the flourishing of all the interrelated goods, including the one here and now destroyed or impeded, within the human community. He illustrates this by suggesting that contraception can be justified in impeding the good of procreation here and now when doing this contributes to marriage and family life, goods necessary for fostering the good of procreation itself ([53], pp. 426–428).[41]

This response is simply not adequate. What it comes to is saying that although there is no criterion whereby we can commensurate the human goods at stake in moral choice, we nonetheless do succeed somehow in commensurating them by interrelating or associating them. McCormick admits as much, for he speaks of assessing the greater good as a "prudent bet" and of commensurating in "fear and trembling" ([52], p. 227).[42] Here McCormick acknowledges that we commensurate by *choosing*; yet the problem is to determine norms that are to govern choices and to make them good. His further claim that the choice to destroy or impede one aspect of human flourishing (one human good) is *necessary* here and now in order to protect and foster *all* of the interrelated goods, including the one deliberately destroyed or impeded, is simply asserted, not demonstrated.

It should also be noted that McCormick is the only proportionalist among contemporary Catholic writers defending this position who explicitly admits – and then in response to objections – that it is not possible to commensurate human (non-moral, ontic) goods in order to determine which are "greater" and which "lesser". Yet he previously presented the position as if it entailed such commensuration, and other advocates of the position continue to do so.[43]

A second response to the criticism that proportionalism entails an impossible commensurating of human goods is that the same kind of commensuration is demanded by the 'principle of double effect' (cf. its fourth condition), which the critics of proportionalism accept. Thus McCormick writes that Catholic critics of proportionalism or consequentialism cannot

> avoid the kind of consequentialistic [proportionalistic] reasoning that our sensibilities seem to demand in such conflict cases. For if a good like life is simply incommensurable with other goods, what do we mean by a proportionate reason where death is . . . indirect? Proportionate to what? If some goods are to be preferred to life itself, then we have compared life with these goods. And if this is proper, then life can be weighed up against other values too, even very basic values ([51], p. 48).

Here it is necessary to note that advocates of proportionalism claim that one can determine the *moral norm* or standard regulating human choice and action by commensurating human goods. Those who reject proportionalism and accept the 'principle of double effect' deny this claim. In articulating this principle, language that sounds proportionalistic was employed in expressing the fourth condition of the principle, but this does not mean that those using this language assumed, as the proportionalists do, that one can determine a moral norm by commensurating human goods and then choosing the act that will result in the 'greater good'. Catholic writers who accept the 'principle of double effect' and reject proportionalism hold that a moral norm or standard has *already* been established, namely the norm that one ought not directly to intend evil and that one ought not to choose to do evil so that good may come about (cf. the second and third conditions of the 'principle of double effect'). Within the context provided by these moral norms, a normative, moral meaning is already presupposed by expressions implying a comparison of goods. Yet in view of the possibility that language implying a comparison of goods may lead to proportionalism, namely, the position that one can determine what *morally* ought to be done by commensurating non-moral (ontic) human goods and choosing that act that will achieve the 'greater (ontic) good,' Germain G. Grisez is correct in saying that it is better to express the fourth condition of the 'principle of double effect' differently, by making it clear that this condition excludes the violation of any other relevant mode of moral responsibility when one chooses to do something which will have bad consequences. The mode of responsibility which forbids doing evil to achieve good is protected by the second condition of the principle, but there are ways other than by deliberately choosing to impede or destroy basic human goods that we can act immorally with respect to them, and these ways are excluded by the fourth condition of the principle ([29], pp. 49–62). For example, one can unfairly accept bad consequences which a fair person would take care to avoid.

Other objections against proportionalism have been made, and among them the more substantive are those charging that this mode of moral reasoning is rooted in a dualistic understanding of the human person and is a form of extrinsicism with respect to the significance of human acts. It is not possible here to develop these criticisms in detail, but a word of explanation is necessary. Writers (e.g., Theo Belmans [4], pp. 327–411, Germain G. Grisez [28], and William E. May [49]), who advance this criticism (dualism) note that proportionalists, in comparing human goods, claim that goods dependent upon conscious awareness for their realization (e.g., goods such as friendship and interpersonal relationships such as the communicative good of marital union) are 'greater' than goods of human persons (such as bodily life and the capacity to engender human life) whose realization does not require conscious awareness. Therefore, the proportionalists hold, such 'lower' goods can be impeded or destroyed should their continued flourishing within the human person inhibit participation in the 'higher', consciously experienced goods. The critics argue that this implies that only subjects consciously aware of themselves as selves truly count as human persons and that some living human beings are no longer persons.

The criticism that proportionalism is a kind of extrinsicism is based on the fact that for the proportionalists no human choice and act is *morally* evil because it is the sort or kind of choice and act that it is (e.g., the choice deliberately and with direct intent to kill an innocent human being) but is *morally* evil only if there is no proportionate good, extrinsic to the deed itself, capable of serving as a proportionate reason for the choice to do evil.[44] When this criticism was raised earlier in the debate, McCormick reacted by saying that "it no longer serves the purposes of constructive moral discourse to argue" in this way ([55], pp. 708–709). Yet, surely it is not inappropriate to note that proportionalism is a form of extrinsicism. For the proportionalists, as has been seen and as McCormick acknowledges ([55], pp. 713–717), hold that *no* human choices and acts are *intrinsically* wicked in a *moral* sense. Their common claim is that we may rightfully choose, with set purpose and with direct, deliberate intent, to do what is *intrinsically* evil in a weak sense so that 'greater good' may come about. They hold that "it is the presence or absence of a proportionate reason" – a good extrinsic to the act chosen and done – "which determines whether my action" is *morally* proper or reprehensible ([51], p. 53). Thus to claim that proportionalism entails a form of extrinsicism does not, in my opinion, destroy the purposes of constructive moral discourse.

To make this claim, however, does not show that proportionalists are mistaken in their position. But here we come to the nub of the debate currently taking place among Catholic writers. Many proportionalists hold that the direct-indirect distinction is not of *moral* significance and that we can choose to do evil for the sake of good to come. For them, 'beneficence' entails not only the requirement to do good but also the requirement to assess the consequences of our choices and acts and to do the deed that will bring about the most good and the least evil or harm, a position that is surely congenial to that held by modern philosophers who invoke the 'principle of utility' in seeking to determine what constitutes 'beneficent' action, no matter how strenuously Roman Catholic proportionalists seek to distance their position from 'utility'.

Those Catholic writers who reject proportionalism hold that the direct-indirect distinction is of crucial moral significance, and they do so because they hold that human choices and acts are not physical events in the world but are rather constitutive of the moral being of the human person. In and through our choices and deeds we determine our selves, our being. In making good moral choices we are, these writers hold, to make those choices in accordance with the fundamental requirements or principles of practical reasonableness or of 'natural law'.[45] The very first of these principles is that *good is to be done and pursued and evil is to be avoided*, and the 'good' in question, as Thomas Aquinas makes quite clear in his discussion of the principle ([1], 1–2, q. 94, a. 2), is not restricted to what proportionalists call the 'moral' good in contradistinction to the 'premoral' or 'ontic' good. The 'good' in question means every true human good, every real good of human persons. A proper understanding of this basic requirement of practical reasonableness shows the proposal that one can rightfully *intend that evil be done* is a proposal that contradicts the most fundamental requirement of practical reasonableness or natural law, inasmuch as this proposal holds that *evil is to be done and pursued*, whenever this is necessary in order to achieve some more 'proportionate' or 'greater' good.

Catholic writers who oppose proportionalism and hold that we are not freely to choose to do evil for the sake of any alleged 'greater' good maintain that we are to be 'beneficent'. But they refuse to accept as 'beneficent' those human choices and acts that have, as their precise 'intent', the destruction or impeding of real goods of real human persons. They hold that such human choices and actions not only do evil, in the sense of harming human persons (e.g., killing innocent human persons), but also, as Vatican II affirms, "debase the perpetrators more than the victims" ([78], n. 27, p. 928). Such actions and choices do so inasmuch as through and in them human persons freely determine themselves to *be* evildoers; they do so freely even if reluctantly and regretfully.

The direct-indirect distinction, these writers hold, is of crucial moral significance precisely because it is in and through our choices, *through what we intend or propose freely to do*, that we relate our moral selves to the real goods of human persons and to the *Summum Bonum*, God, who is the source of these goods, their ultimate giver. These writers hold that a proper moral attitude, one required both by the basic principles of practical reasonableness or 'natural law' and by God's reign of holiness and love, is one of openness to and reverence for the basic goods of human persons and of the persons in whom they are meant to flourish. One way of violating this moral attitude is to choose, freely and with direct intent, to impede, damage, or destroy anything truly good, deeming it of 'less' importance than some other incommensurable human good that we may happen to 'prefer' here and now.[46]

In this section I have sought to describe the principal issues being debated by contemporary Roman Catholic writers. In particular, I have sought to describe the development of the notion of 'proportionalism' by some Catholics, with its attendant understanding of 'beneficence', and the reasons why other Catholic writers repudiate this development and judge it and its correlative understanding of 'beneficence' to be erroneous. Those Catholic writers who reject proportionalism hold that the specific moral norms authoritatively proposed by the magisterium are capable of being shown to be true by relating those norms to the "universally binding principles of divine and natural law". They believe that by freely shaping our lives in accord with these norms we will in truth choose to act 'beneficently'.

IV. CONCLUDING REFLECTIONS

As a nonproportionalist, nonconsequentialist Roman Catholic, I hold that the norms, both universal and more specific, proposed by the teaching authority of the Church are true. 'Beneficent' deeds are those in which we affirm the real goods of human persons and respect them. No deed is 'beneficent' in which we freely choose to do evil, that is, to impede, prevent, or destroy a good meant to flourish in a human person. With respect to health care this means that we ought, in our choices and actions, to respect the goods of human persons, goods such as life itself, human procreativity or the power to give life, and justice, which requires us to regard others as we would have them regard us, that is, as irreplaceable persons and not as substitutable and replaceable individuals. On this understanding of 'beneficence', no choice to destroy human life, whether our own or others, whether unborn or senile, can be regarded as 'beneficent'. To choose to kill is not to choose to do good but to do evil; it is to choose death, not life. 'Beneficent' euthanasia, to use an example, whether chosen for the unborn, the newborn, or the dying, is a contradiction camouflaged by a euphemism, for the choice to kill is a choice to destroy, not respect, the good of human life. If one reflects seriously on other issues in modern health care, one will come to realize that no choices executing proposals directly and precisely intended to destroy, impede, or prevent the flourishing of human goods within human persons can be considered 'beneficent'. "Good is to be done and pursued; evil is to be avoided." If this principle inwardly shapes our choices and actions we will be 'beneficent' and human persons and human communities will flourish. If the pseudo-principle that "evil is to be done for the sake of 'greater' good to come" is adopted, health care may soon be reduced to an analysis of economic costs and benefits and irreplacable human persons will be reduced to substitutable individuals.

The Catholic University of America
Washington, D.C.

NOTES

[1] The teaching office or magisterium of the Pope and of the Bishops throughout the world in communion with him is affirmed by the Roman Catholic Church as an agency willed by Christ himself. On this, see [79].

[2] It should be noted that while Aquinas affirms that while the very first precept or principle of natural law is that *good is to be done and pursued and evil is to be avoided*, he insists that there are several 'first' precepts or principles of natural law ([1], 1–2, q. 94, a. 2). An excellent commentary on this teaching of Aquinas and its importance for ethical theory is provided in [25].

[3] This teaching is constantly affirmed by the magisterium. Among the principal sources in which this teaching is found are: [38], p. 593; [63], p. 423; [77], nn. 2–3, pp. 800–801; and [78], n. 16, p. 916 and n. 51, p. 955.
[4] With respect to the teaching of the magisterium summarized in the propositions given in the text, see [19].
[5] Also see [77], nn. 1–2, pp. 799–801.
[6] On this, see [76], n. 7, pp. 753–754. Here it should be noted that the constant teaching of the Roman Catholic Church is that moral truth about *some* moral matters *can* be securely known without the help of divine revelation, but that moral truth about other matters can be *securely* known only because God has revealed what objectively ought to be done and has given to His Church the mission of teaching this truth. On this, see Vatican Council I, *Dei Filius*, c. 2 in ([13], nn. 3004–3007, p. 588). The teaching of Vatican Council I was clearly reaffirmed by Vatican Council II; see [77], especially nn. 3 and 14, pp. 801–802 and 810–811.
[7] See, for example, [78], n. 16, p. 916.
[8] See also [58], p. 9.
[9] For teaching on abortion, see [64], [68], and [78], n. 27, p. 928. For teaching on euthanasia, see [70] and [78], n. 27, p. 928.
[10] For teaching on contraception and contraceptive sterilization, see [33], pp. 278–280; [61], n. 14, pp. 337–338; [62], pp. 41–43; and [69].
[11] [61], n. 14, pp. 337–338, similarly stresses that the Church condemns *direct* abortion.
[12] All sources listed in notes 9 through 11 repeatedly affirm that the acts described violate the requirements of the natural and divine law.
[13] Documents of the Church's magisterium in which an appeal is made to the 'principle of totality' include [62], p. 46 and [65], p. 779.
[14] The meaning of the 'principle of totality' as understood by the magisterium is set forth in more detail in [66], p. 673.
[15] See also [12], p. 211 and [69].
[16] See [3], p. 136.
[17] Thus Charles E. Curran, in his Foreword to O'Connell's book, and Joseph Fuchs, S. J., a noted German theologian, in his comments on the jacket of the book.
[18] The term proportionalism has now become widely accepted as a way of designating this current of thought in contemporary Roman Catholic moral thinking. Thus, Paul E. McKeever, who is himself a proponent of the position, uses this term to describe the position in [56].
[19] Among the leading advocates of proportionalism among Roman Catholic writers are Joseph Fuchs, Bruno Schuller, Louis Janssens, Franz Scholz, Richard A. McCormick, Daniel Maguire, Philip Keane, and the authors of a report on human sexuality commissioned by the Catholic Theological Society of America (Anthony Kosnik, Agnes Cunningham, William Carroll, James Modras, and James Schulte). Charles E. Curran also accepts the principle of proportionate good and assents to the proposition given in the text, a proposition affirmed in the writings of the theologians named here. Key articles by Fuchs, Schuller, Janssens, Scholz, McCormick, and Curran have been gathered and published in [11]. The development of proportionalism has been well described by McCormick in [51] and [55].

Maguire's presentation of proportionalism is found in [41], pp. 77–130. Keane's position is found in [34], pp. 43–51. The position of Kosnik, *et al.* is presented in [37], pp. 88–98.

20 Readers should know that other Catholic writers challenge this position. See in particular [20]. Of central importance here is a passage in [79], n. 25, pp. 379–380.

21 See [55], pp. 652–668 for a discussion of this matter. See also the "Pastoral Guidelines" in [37], pp. 99–239, guidelines for practice that quite frequently support positions contrary to the specific teachings of the magisterium.

22 This position is widely held. For contraception, a representative sampling is given in [9]. For sterilization, see [10], pp. 194–211 and [53], pp. 260–280.

23 For instance, see [12], pp. 109–136; [41], pp. 199–202; and [10], pp. 171–193.

24 See [41], along with [42].

25 As McCormick puts it in [51], p. 53, "it is the presence or absence of a proportionate reason which determines whether my action . . . involves me in turning against a basic good in a way which is *morally* reprehensible" (emphasis added).

26 This is the usual way of stating the principle. See, for instance, [6]. For a history of this principle, see [23] and [43].

27 See, for instance, [16] and [35].

28 An important early article reopening discussion of the 'principle of double effect' was that of Peter Knauer in [36]. An incisive criticism of Knauer's essay (along with essays by William Van der Marck and Cornelius Van der Poel) was made by Germain G. Grisez in [27].

29 [74], pp. 41–70 and in particular pp. 54–61. A good analysis of Van der Marck is provided both by Grisez in [27] and by McCormick in [51], pp. 12–24.

30 See [75]. Again Grisez [27] and McCormick [51] provide good analyses of Van der Poel.

31 But compare this treatment with McCormick's later understanding of the validity of the 'direct-indirect intent' distinction in his essay in [52], pp. 254–262. In [51] (which is reprinted as chapter one in [52]), McCormick had attributed some moral and not merely descriptive import to the distinction; in the latter work it has only a *descriptive* import when the evil in question is 'premoral' as opposed to 'moral'.

32 For this, see works cited in note 19 and [42].

33 Albert Di Ianni makes clear this distinction between 'intrinsic evil in a weak sense' and 'intrinsic evil in a strong sense' in his essay in [14]. McCormick accepts this distinction; see [55], pp. 713–717.

34 This is more clearly affirmed by McCormick in [54].

35 Among non-Catholic religious ethicists who oppose this development within Roman Catholic thought are Frederick S. Carney [5] and Paul Ramsey [52], pp. 69–154. A secular author who sharply criticizes this type of thinking is Alan Donagan in [15], Chapter 6.

36 Among Catholic writers who have strongly criticized the development of 'proportionalism' are Germain G. Grisez, ([26], [27], [28], and [29]; John Finnis, ([17], [18], and [19]); Theo Belmans, [4]; Joseph Boyle, [30]; John Connery, [7]; Paul Quay [67], and William E. May ([44], [47], and [48]).

37 On the initial plausibility of the proportionalist proposal, see in particular [30], Chapter 11.

38 This is the point most sharply pressed by Ramsey, Grisez and Boyle. See works cited in notes 35 and 36.

[39] Here I am attempting to summarize, all too briefly and with insufficent explanation, the thought underlying those Catholic writers who oppose proportionalism. This thought is developed in much more detail in [17], Part Two; [24], Chapter 3; [27]; [30], Chapter 11; and [44], Chapters 2, 3, and 4.

[40] Grisez argues that the proposal is ultimately meaningless; see [29].

[41] McCormick discusses at more length the 'interrelatedness' of human goods in [52], pp. 233–241 and 251–254;

[42] "*Somehow or other*, in fear and trembling, we commensurate" (italics in original). See also p. 230. "I would suggest that . . . the means of assessing the lesser evil [is] a prudent bet if you will."

[43] Thus O'Connell insists that what constitutes right action is "that action which contains the proportionally greatest *maximization* of good and *minimization* of evil" ([60], p. 153, emphasis added).

[44] This criticism was initially made in [45]. It was developed in [47] and in [48].

[45] Grisez and Finnis usually use the expression "requirement of practical reasonableness" in place of "requirements of the natural law," because of the many meanings that have been attached to 'natural law'. The position is developed by Grisez in [27] and with Joseph Boyle in [30], and by Finnis in [17].

Grisez ([24], Chapter 3) provides an excellent critique of the "conventional natural law theory", rooted in the thought of Francis Suarez and reflected in many manuals of moral theology, as one that seeks to determine moral obligations from factual considerations, and hence committing the fallacy of deriving an *ought* from an *is*. He shows how this theory differs profoundly from the natural law theory set forth by Thomas Aquinas ([1], 1–2, qq. 90–97; also see [46]). Yet Grisez, and with him Finnis, argue that Aquinas proceeded too abruptly to derive specific moral norms from the first precepts or principles of 'natural law'. They argue that the first principles of 'natural law' are principles or requirements of 'practical reasonableness' and that it is necessary to develop various 'modes of responsibility' demanded by these 'principles of practical reasonableness' prior to determining specific normative rules. See [17], [19], and [30], Chapter 11.

[46] On the 'modes of responsibility' see works cited in note 45.

BIBLIOGRAPHY

1. Aquinas, T.: 1948, *Summa Theologiae*, 4 vol., Marietta, Rome.
2. Augustine: 1952, *Contra Mendacium*, C. 6. 15, (transl. H. B. Jaffe), in Roy J. Deferrari (ed.), *St Augustine: Treatises on Various Subjects*, Fathers of the Church, Inc., New York, p. 141.
3. Beauchamp, T. and Childress, J.: 1979, *Principles of Biomedical Ethics*, Oxford University Press, New York.
4. Belmans, T. G.: 1980, *Le Sens Objectif de l'Agir Humain* (Studi Tomistici 8), Libreria Editrice Vaticana, Vatican City.
5. Carney, F. S.: 1978, 'McCormick on Teleology', *Journal of Religious Ethics* 6, 81–120.
6. Connell, F.: 1968, 'Double Effect, Principle of', *New Catholic Encyclopedia*, McGraw-Hill Co., New York, 4, 1020–1022.

7. Connery, J.: 1979, 'Morality of Consequences: A Critical Appraisal', in C. E. Curran and R. A. McCormick (eds.), *Readings in Moral Theology No. 1: Moral Norms and Catholic Tradition*, Paulist Press, New York, pp. 244–266.
8. Curran, C. (ed.): 1968, *Absolutes in Moral Theology?*, Corpus Books, Washington, D.C.
9. Curran, C. (ed.): 1969, *Contraception: Authority and Dissent*, Herder and Herder, New York.
10. Curran, C.: 1976, *New Perspectives in Moral Theology*, University of Notre Dame Press, Notre Dame, Ind.
11. Curran, C. and McCormick R. A. (eds.): 1979, *Readings in Moral Theology No. 1: Moral Norms and Catholic Tradition*, Paulist Press, New York.
12. Dedek, J: 1975, *Contemporary Medical Ethics*, Sheed and Ward, Inc., New York.
13. Denzinger, H., and Schonmetzer, A. (eds.): 1963, *Enchiridion Symbolorum Definitionum et Declarationum de rebus fidei et morum*, 32nd ed., Herder, Rome.
14. Di Ianni, A.: 1979, 'The Direct/Indirect Distinction in Morals', in C. E. Curran and R. A. McCormick (eds.), *Readings in Moral Theology No. 1: Moral Norms and Catholic Tradition*, Paulist Press, New York, pp. 215–243.
15. Donagan, A.: 1976, *A Theory of Morality*, University of Chicago Press, Chicago.
16. Farraher, J.: 1963, 'Notes on Moral Theology', *Theological Studies* **21**, 69–79.
17. Finnis, J. M.: 1980, *Natural Law and Natural Rights*, Clarendon Law Series, Oxford University Press, Oxford.
18. Finnis, J. M.: 1980, 'Reflections on an Essay in Christian Ethics, Part I and II', *Clergy Review* **65**, 51–57, 87–93.
19. Finnis, J. M.: 1981, 'The Natural Law, Objective Morality, and Vatican II', in W. E. May (ed.), *Principles of the Catholic Moral Life*, Franciscan Herald Press, Chicago, pp. 113–149.
20. Ford, J. C. and Grisez, G. G.: 1978, 'Contraception and the Infallibility of the Ordinary Magisterium', *Theological Studies* **39**, 258–312.
21. Frankena, W. K.: 1973, *Ethics*, 2nd ed., Prentice-Hall, Inc., Englewood Cliffs, N.J.
22. Fuchs, J.: 1979, 'The Absoluteness of Moral Terms', in C. E. Curran and R. A. McCormick (eds.), *Readings in Moral Theology No. 1: Moral Norms and Catholic Tradition*, Paulist Press, New York, pp. 94–137.
23. Ghoos, J.: 1951, 'L 'Acte a Double Effect: Etude de Theologie Positive', *Ephemerides Theologicae Lovaniensis* **27**, 30–52.
24. Grisez, G. G.: 1964, *Contraception and the Natural Law*, The Bruce Publishing Co., Milwaukee.
25. Grisez, G. G.: 1965, 'The First Principle of Practical Reason: A Commentary on the *Summa Theologiae* 1–2, Q. 94, a. 2', *Natural Law Forum* **10**, 168–196.
26. Grisez, G. G.: 1970, *Abortion: The Myths, the Realities, and the Arguments*, Corpus Books, New York.
27. Grisez, G. G.: 1970, 'Toward a Consistent Natural Law Ethics of Killing', *American Journal of Jurisprudence* **15**, 64–97.
28. Grisez, G. G.: 1977, 'Dualism and the New Morality', in *L'Agire Morale* (papers delivered on the 7th Centenary of the Deaths of Thomas Aquinas and Bonaventure), Editrici Domenicani, Naples, pp. 323–330.
29. Grisez, G. G.: 1978, 'Against Consequentialism', *American Journal of Jurisprudence* **23**, 21–72.

30. Grisez, G. G. and Boyle, J.: 1978, *Life and Death With Liberty and Justice: A Contribution to the Euthanasia Debate*, University of Notre Dame Press, Notre Dame, Ind.
31. Grisez, G. G. and Ford, J.: 1978, 'Contraception and the Infallibility of the Ordinary Magisterium', *Theological Studies* **39**, 258–312.
32. Janssens, L.: 1979, 'Ontic Evil and Moral Evil', in C. E. Curran and R. A. McCormick (eds.), *Readings in Moral Theology No. 1: Moral Norms and Catholic Tradition*, Paulist Press, New York, pp. 40–93.
33. John Paul II, Pope: 1979, '"Stand Up" for Human Life', *Origins: NC Documentary Service* **9.18** (October 18), pp. 278–281.
34. Keane, P. S.: 1977, *Sexual Morality: A Catholic Perspective*, Paulist Press, New York.
35. Kelly, G.: 1952, 'Notes on Moral Theology', *Theological Studies* **13**, 59–61.
36. Knauer, P.: 1979, 'The Hermeneutic Function of the Principle of Double Effect', in C. E. Curran and R. A. McCormick (eds.), *Readings in Moral Theology No. 1: Moral Norms and Catholic Tradition*, Paulist Press, New York, pp. 1–39.
37. Kosnik, A., *et al.*: 1977, *Human Sexuality: New Directions in American Catholic Thought*, Paulist Press, New York.
38. Leo XIII, Pope: 1888, Encyclical *Libertas Praestantissimum*, in *Acta Sanctae Sedis* **20**, 593–613.
39. Liebard, O. (ed.): 1978, *Official Catholic Teachings: Love and Sexuality*, Consortium, Wilmington, N.C.
40. Maguire, D. C.: 1968, 'Moral Absolutes and the Magisterium', in C. E. Curran (ed.), *Absolutes in Moral Theology?*, Corpus, Washington, pp. 57–107.
41. Maguire, D. C.: 1974, *Death by Choice*, Doubleday, New York.
42. Maguire, D. C.: 1975, 'A Catholic View of Mercy Killing', in M. Kohl (ed.), *Beneficent Euthanasia*, Prometheus, Buffalo, N.Y., pp. 34–43.
43. Managan, J.: 1949, 'An Historical Analysis of the Principle of Double Effect', *Theological Studies* **10**, 41–61.
44. May, W. E.: 1975, *Becoming Human: An Invitation to Christian Ethics*, Pflaum, Dayton.
45. May, W. E.: 1977, 'Contraception, Abstinence, and Responsible Parenthood', *Faith and Reason* **3**, 34–52.
46. May, W. E.: 1977, 'The Meaning and Nature of Natural Law in Thomas Aquinas', *American Journal of Jurisprudence* **22**, 168–189.
47. May, W. E.: 1978, 'Modern Catholic Ethics: The New Situationism', *Faith and Reason* **4**, 21–38.
48. May, W. E.: 1978, 'The Moral Meaning of Human Acts', *Homiletic and Pastoral Review* **79** (October), 10–21.
49. May, W. E.: 1979, 'Toward an Integrist Understanding', in Dennis Doherty (ed.), *Dimensions of Human Sexuality*, Doubleday, New York, pp. 95–124.
50. May, W. E. (ed.): 1981, *Principles of Catholic Moral Life*, Franciscan Herald Press, Chicago, pp. 113–149.
51. McCormick, R. A.: 1973, *Ambiguity in Moral Choice*, Marquette University Dept. of Theology, Milwaukee.
52. McCormick, R. and Ramsey, P. (eds.): 1978, *Doing Evil to Achieve Good: Moral Choice in Conflict Situations*, Loyola University Press, Chicago.

53. McCormick, R. A.: 1981, *How Brave a New World? Dilemmas in Bioethics*, Doubleday, New York.
54. McCormick, R. A.: 1981, 'Letter to the Editor', *Hospital Progress* **62** (January), 6.
55. McCormick, R. A.: 1981, *Notes on Moral Theology 1965 through 1980*, University Press of America, Washington, D.C.
56. McKeever, P.: 1981, 'Proportionalism as a Methodology in Catholic Moral Teaching', in *Human Sexuality and Personhood*, Pope John Center, St. Louis, pp. 211–222.
57. National Conference of Catholic Bishops: 1973, *Ethical and Religious Directives for Catholic Health Care Facilities*, United States Catholic Conference, Washington, D.C.
58. National Conference of Catholic Bishops: 1976, *To Live as Christ Jesus: A Pastoral Reflection on the Moral Life*, United States Catholic Conference, Washington, D.C.
59. Nassali-Rocca, E.: 1967, 'Hospitals, History of: The Christian Hospital to 1500', *New Catholic Encyclopedia*, McGraw-Hill, New York, 7, 159–163.
60. O'Connell, T. E.: 1978, *Principles for a Catholic Morality*, Seabury Press, New York.
61. Paul VI, Pope: 1968, Encyclical *Humanae Vitae*, in O. M. Liebard (ed.), *Official Catholic Teachings: Love and Sexuality*, Consortium, 1978, Wilmington, N.C., pp. 331–347.
62. Pius XI, Pope: 1930, Encyclical *Casti Connubii*, in O. M. Liebard (ed.), *Official Catholic Teachings: Love and Sexuality*, Consortium, 1978, Wilmington, N.C., pp. 23–70.
63. Pius XII, Pope: 1939, Encyclical *Summi Pontificatus*, in *Acta Apostolicae Sedis* **31**, 413–453.
64. Pius XII, Pope: 1945, 'Address to Italian Medical Guild of St. Luke: On the Moral and Social Duties of the Medical Profession', in *Discorsi e Radiomessaggi di Sua Santita Pio XII*, Vatican City Press, **6**, 183–195.
65. Pius XII, Pope: 1952, 'Address to the Delegates to the First International Congress on the Histopathology of the Nervous System: On the Moral Limits of the Medical Research and Experimentation', *Acta Apostolicae Sedis* **44**, 779–787.
66. Pius XII, Pope: 1953, 'Address to the Twenty-Sixth Convention of the Italian Society of Urology: On the Excision of a Healthy Organ, Especially Castration for Cancer', *Acta Apostolicae Sedis* **45**, 673–682.
67. Quay, P.: 1979, 'Morality by Calculation of Values', in C. E. Curran and R. A. McCormick (eds.), *Readings in Moral Theology No. 1: Moral Norms and Catholic Tradition*, Paulist Press, New York, pp. 267–293.
68. Sacred Congregation for the Doctrine of the Faith: 1974, *Declaration on Procured Abortion* (18 Nov.), United States Catholic Conference, Washington, D.C.
69. Sacred Congregation for the Doctrine of the Faith: 1975, *Declaration on Sterilization* (13 March), in *Origins; NC Documentary Services* **6.3** (10 June 1976), 33–35.
70. Sacred Congregation for the Doctrine of the Faith: 1980, *Declaration on Euthanasia* (5 May), United States Catholic Conference, Washington, D.C.
71. Scholz, F.: 1979, 'Problems on Norms Raised by Ethical Borderline Situations: Beginnings of a Solution in Thomas Aquinas and Bonaventure', in C. E. Curran and R. A. McCormick (eds.), *Readings in Moral Theology No. 1: Moral Norms and Catholic Tradition*, Paulist Press, New York, pp. 158–183.
72. Schuller, B.: 1979, 'Direct Killing, Indirect Killing', in C. E. Curran and R. A. McCormick (eds.), *Readings in Moral Theology No. 1: Moral Norms and Catholic Tradition*, Paulist Press, New York, pp. 138–157.

73. Schuller, B.: 1979, 'Various Types of Grounding for Ethical Norms', in C. E. Curran and R. A. McCormick (eds.), *Readings in Moral Theology No. 1: Moral Norms and Catholic Tradition*, Paulist Press, New York, pp. 184–198.
74. Van der Marck, W.: 1967, *Toward a Christian Ethic: A Renewal in Moral Theology*, Newman Press, Westminster, Md.
75. Van der Poel, C.: 1968, 'The Principle of Double Effect', in C. E. Curran (ed.), *Absolutes in Moral Theology?* Corpus, Washington, D. C., pp. 186–210.
76. Vatican Council II: 1963–1965, '*Dei Verbum*' (Dogmatic Constitution on Divine Revelation), in Austin Flannery (ed.), *Vatican Council II: The Conciliar and Post Conciliar Documents*, Costello Publ. Co., pp. 750–765.
77. Vatican Council II: 1963–1965, '*Dignitatis Humanae*' (Declaration on Religious Liberty), in Austin Flannery (ed.), *Vatican Council II: The Conciliar and Post Conciliar Documents*, Costello Publ. Co., Northport, N.Y., 1975, pp. 799–812.
78. Vatican Council II: 1963–1965, '*Gaudium et Spes*' (Pastoral Constitution on the Church in the Modern World), in A. Flannery (ed.), *Vatican Council II: The Conciliar and Post Conciliar Documents*, Costello Publishing Co., Northport, N.Y., 1975, pp. 903–1001.
79. Vatican Council II: 1963–1965, '*Lumen Gentium*' (Dogmatic Constitution on the Church), in A. Flannery (ed.), *Vatican Council II: The Conciliar and Post Conciliar Documents*, Costello Publishing Co., Northport, N.Y., 1975, pp. 350–426.

HARMON L. SMITH

PROTESTANT ETHICS AND BENEFICENCE[1]

In the late 1960's, before matters of medicine and ethics became public and popular, I listened to an English doctor who stated unequivocally before an international conference that theologians are no longer (were they once!) of any help to physicians and surgeons, and that in consequence doctors themselves must formulate "their own ethical code" which will both satisfy the majority of doctors and (he added confidently) the well-being of patients and the best interests of society as well. In the intervening years, in hospitals and medical schools both here and abroad, I have many times encountered a similar sentiment, located most often in an autobiographical account of how, as an undergraduate medical student or young house officer, one had sought help from a chaplain or pastor or academic theologian, only to learn that "he had nothing useful to tell me."

I cannot say certainly that this mood among physicians is predominant or prevailing, but it is surely not infrequent or unknown; and my own experience with physicians confirms that the damaging ineptitude of religious professionals has made many doctors reluctant or indifferent or opposed to seeking out the advice and assistance of theologians. In the process, they have also developed efficient, if not elaborate, ways to separate their professional practice from religious consciousness.[2]

Why medical professionals might be disposed to seek the help of theologians in ethically significant situations should not be at all mystifying if we remember that Western culture is *de facto* Judeo-Christian; which means only that the institutions and professions of Western culture participate in a common heritage which makes them culturally *de facto* Judeo-Christians. I do not mean, of course, that all of us affirm the same doctrines or assent to the same creeds; but our ethical quandaries *do* occur in a particular kind of culture, which operates with certain discrete views of man and the world, and which can be properly called Judeo-Christian. When those conflicts and quandaries are severed from the 'point-of-view' which alone makes them intelligible, however, it is little wonder that theologians are no longer of any help to physicians and surgeons. Indeed, in seeking our commonality elsewhere than our shared cultural heritage, none of us can be of much help to the other. Paul Ramsey has put the matter in a splendid epigram: we cannot agree upon

Earl E. Shelp (ed.), Beneficence and Health Care, 153–182.

the practice of virtue, he said, because we will not agree upon the principles of virtue.

The history of the relationship between Western religious traditions and Western medicine shows both cooperation and antagonism. In recent decades, our scientific and technical achievements have (more often than not) outstripped not only the capacities of positive law (our public morals) but our moral imagination as well. So our society now finds itself in the somewhat anomalous circumstance of being adversarial about both means and ends in controversies ranging from genetic intervention to the delivery of health care to management of the manner and time of human dying and death. If it is so that we share a certain cultural heritage, how can this be? How could this condition have developed? Is it possible any longer to inquire intelligently of the One among the Many?

Since the Protestant Reformation and the modern scientific era emerge in a roughly coincident (some say, reinforcing) historical period, and since both the religious and medical institutions of Western culture are committed to serving fellowman by doing good and charitable acts, it may be instructive to examine one of these (*viz.*, Protestant Christianity) with a view toward assessing its role in the evolution of our present situation.

There are several evident risks when dealing with a religious phenomenon as prolix and diffuse as Protestantism. It is arguable that a thorough treatment of Protestant ethics and beneficence ought to include not only the conventionally classical Protestant traditions but their diverse derivative offspring as well. One might take many starting places and avenues of development. But to undertake examination of the many varieties of modern Protestant theological ethics would make it quickly evident that methodological problems alone are prohibitively numerous and severe (especially in so brief a space as this). I propose, therefore, to consider briefly the theological and ethical lineaments of three principal figures (Martin Luther, John Calvin, and John Wesley) who exercised primary formative influences on Protestant Christianity in the 16th and 18th centuries and who are representative of major traditions in Protestantism. Thereafter I will offer some critical observations on the recent and contemporary Protestant *Weltanschauungen* as regards ethics and beneficence. I have elected this course because Luther, Calvin, and Wesley are prominent among the fountainhead sources to which much of modern mainline Protestantism appeals for theological authority. First, however, we must set the stage.

Despite a series of persistent discordant notes and struggles, both internal and external, Christian orthodoxy in the West enjoyed relative stability and solidarity through the 13th century. In a way that has not been iterated since, medievalism represented the ideal of a single and united human family under the benevolent direction of Christ's vicar as its father and head. Although that often represented more an ideological than an actual historical reality, the vision of constancy and harmony was nevertheless pervasive. The great synthesizing work of Thomas Aquinas (d. 1274) both reinforced the regnant medieval belief (that religion is rational and reason divine) and made that belief credible by demonstrating that all knowledge and truth are harmonious.

But this era was moving toward its end. The church – and especially the papacy – was becoming increasingly secularized and was almost constantly under attack; the gradual disintegration of the Holy Roman Empire was producing far-reaching political results; and the social and economic foundations of feudalism were weakening owing to crop failures and famines, perennial armed conflict, and the profound effects of the Black Death (1347 –1351) which not only decimated populations (including the ranks of the clergy) but, perhaps more importantly, was widely interpreted as a direct expression of divine wrath. Meantime, the medieval majesty accorded reason was embraced by advocates whose primary interest was other than the defense and exaltation of the religious and cultural status quo; accordingly, increasingly numbers began turning away from intellectualism to mysticism and from supernaturalism to humanism.

In order to comprehend, if only in the most rudimentary way, the lineaments of modern Protestant Christianity, it is necessary to appreciate the coincident emergence of two movements in the 13th and 15th centuries, one external to organized religion *qua se* and the other rather more internal, which further contributed to the erosion of the comfortable equilibrium of the medieval synthesis. I have already alluded to one of these; that group of Renaissance thinkers – among them, Petrarch and Boccaccio – who attempted to rediscover the classical traditions of ancient Greece and Rome as an alternative to the religiously dominated culture of medieval Europe. Confident in human capacities and interested in this-worldly affairs, these folk were relatively uninterested in life *after death*, revelation from *beyond unaided reason*, sacraments, theology, and all the rest that are familiar hallmarks of medieval church and society. By the late 14th and early 15th centuries, the Roman Church had begun to associate this Renaissance with religious revolt and had branded it as both heretical and schismatic – which explains, at least partly, why humanists in the north were initially congenial toward the Protestant insurgence. Indeed, it was remarked that Erasmus laid the egg which Luther hatched; and that Luther later observed, "But I expected a different sort of bird!"

Already there had been a series of pre-reformers – among them Wyclif, Huss, and Savonarola – who, in various ways, prefaced the internal conflict which produced Reformation. Begun by Luther (1483–1546) in Germany, reinforced by Calvin (1509–1564) in Geneva, augmented by Cranmer (1489–1556) in England, and vastly extended by more radical movements like Anabaptism and Socinianism, Protestantism found fallow ground in the 16th century ethos. The history of this period sufficiently attests to the violent dislocation of medieval tranquility in every aspect of life – social, political, economic, religious; and while we cannot attend to the full range of these turbulent times, it is nevertheless important to remember them as the setting within which a significantly different set of ideas and beliefs and teachings began to gain ascendancy.

I. LUTHER

Martin Luther's principal theological contributions may be summarized, at the expense of appreciating fully his remarkable breadth and subtlety, in familiar phrases: the Bible is the supreme authority in matters of faith and conduct; justification is by grace alone; and it is each other's duty to be Christ to the other in the priesthood of all believers. As it has turned out, both the formulation and the implementation of these concepts have had enormously far-reaching effect, not only within Protestantism but for reforming movements within Catholicism as well.

Until Luther's declaration of scriptural supremacy, and his subsequent translation of the Bible into German, the laity had been largely deprived of first-hand knowledge of scripture both by its Latin text and a forbidding accumulation of exegetical and hermeneutical glosses and traditions. Now Luther declared that the heart of any believer could be addressed directly, and the entire Bible accordingly judged, by Jesus' message of salvation by grace through faith. To make the Bible available in the vernacular (despite considerable illiteracy), to substitute scriptural authority for that of popes and councils, and to invite individual Christians to interpret the text according to their own piety and experience gave common people an extraordinary freedom in respect of both the old Catholicism and the new Protestantism. Moreover, the gulf between privileged clergy and benighted laity was further spanned by the doctrine of the priesthood of all believers. Over time, both of these tenets would come to be treated far less seriously than Luther himself intended. Meanwhile, in his efforts to break down the walls which protected Rome's power and authority,[3] Luther had sown the seeds which would radically democratize religious authority.

As Biblical supremacy voided papal sovereignty, and the priesthood of believers nullified priestly domestication of sacramental religion, so the doctrine of *justificatio sola gratia* manumitted ordinary people from dependence either upon their own works or the Church's dispensation of indulgence and merit. Good work, beneficence, is possible to be sure; indeed, it is expected; but if it comes, it does so now directly and freely from God as the fruit of grace faithfully received.[4]

Luther's first lectures were on the Psalms (1513–1515), and it is clear that his encounter with the 22nd Psalm (which begins, "My God, my God, why has Thou forsaken me?") confronted him with the quintessential theological question: why should these words be on the lips of Christ upon the cross, what can they mean, why was Christ forsaken? The answer must be that "For our sake he made him to be sin who knew no sin, so that in him we might become the righteousness of God" (2 Cor. 5: 21). Typically, nowadays, 'justification' means the sufficient grounds for defense, usually of a proposition or judgment, as in: 'There is no justification for the Watergate caper.' But in Luther's thought, the term was used very differently; in a word, it meant forgiveness. It did not mean that a sinner was made righteous, nor even that one was accorded the benefit of the doubt! Sinful men and women are just that – sinners; the unrighteous are *accounted* righteous; and God's love toward us is that while we are yet in the condition of sin, without anything to commend us or to merit mercy, we are gifted (χάρις) with pardon, acceptance, reconciliation. Beneficence in Luther's theology is therefore an action reserved to God; it is God's kindly, charitable, gracious gift to us of His love.

Such a view of God's grace, if cogent, requires a certain theological anthropology; and Luther expounded his views in response to Erasmus's tract, *De libero arbitrio*. It had become clear to Erasmus that his principle disagreement with Luther had to do with the doctrine of man. Luther agreed; indeed, in the conclusion of his *De servo arbitrio*, he gave Erasmus " . . . hearty praise and commendation . . . that you alone, in contrast with all others, have attacked the real thing, that is, the essential issue . . . you, and you alone, have seen the hinge on which all turns, and aimed for the vital spot" ([13], p. 786). That 'vital spot' was precisely whether beneficence, or even the possibility of it, is within human capacity; and Luther's answer was an emphatic 'no!' Man's will, wrote Luther, is "like a beast standing between two riders. If God rides, it wills and goes where God wills . . . . If Satan rides, it wills and goes where

Satan wills. *Nor may it choose to which rider it will run, or which it will seek; but the riders themselves fight to decide who shall have and hold it*" ([13], p. 635; italics added). Throughout *De servo arbitrio* Luther is intent upon identifying 'free will' with natural, rational knowledge of God – an enabling cognition rather than a volitional ability – and this accounts for his discussion, early on, of epistemology. The question of free will is not so much concerned with nature-vs-grace as it is with volitional (or even a qualifiedly free) response vs. a mechanistic determinism. Thus the real point is *sola gratia*, but with a twist.

Far from denying *sola gratia*, Erasmus – like Luther – saw God to be the efficient cause of every effect, even secondary causality. He further recognized that in things necessary the *causa deficiens* must be reduced to the *causa efficiens*, and that there is therefore no final distinction between what God himself produces and what He merely allows to come about in consequence of not hindering it. Erasmus defined free will as "a power of the human will, by virtue of which man can apply himself to all that concerns eternal salvation or can turn away from it" ([15], I, B, 10). Luther, on the other hand, held that it is audacious to attempt to harmonize the "wholly free foreknowledge of God" with our own freedom. "Natural reason herself is forced to confess that the living and true God must be One who by His own liberty imposes necessity on us .... Either God makes mistakes in His foreknowledge, and errors in His actions (which is impossible), or else we act, and are caused to act, according to His foreknowledge and action ... we are not made by our own will, but by necessity, and accordingly ... we do not do anything by right of 'free-will,' but according to what God foreknew and works by His infallible and immutable counsel and power" ([13], pp. 718– 719). The fundamental difference between Luther and Erasmus thus evolves about the *relation* of divine grace and human response to divine sovereignty. Erasmus saw no antithesis; Luther did. The result, for Luther, is that theologically warranted possibilities for human beneficence *as such* are extraordinarily limited if not entirely foreclosed; and the ethical product is, understandably, Luther's dualism (or as some insist, paradox) of the two kingdoms, one secular and the other sacred, which are methodologically mutually exclusive.

II. CALVIN

Eleven years after Luther's celebrated debate with Erasmus, John Calvin published the *Institutes of the Christian Religion* (1536), Protestantism's first systematic theology [2]. Between Calvin and Luther there is much in common; but there are also distinctive themes in the *Institutes* which, in the succeeding two centuries, formed theological bedrock for the Protestant dispersion in both Europe and America. Again with all the risks attendant to abbreviation and condensation, a résumé of Calvin's principal theological doctrines can be concisely stated in familiar phrases: the human condition consists of native depravity; God is absolutely supreme and both salvation and condemnation are by His election alone; and the ideal human society consists in a theocratic state.

Descriptions of Calvin's theology customarily give preeminence to the doctrine of divine sovereignty; but, in addition to being somewhat at variance with the development of ideas in the *Institutes*, this priority is internally mistaken on methodological grounds. Calvin's own beginning point is with the human condition: originally endowed "by natural instinct" with "some sense of Deity" ([2], I, iii, 1), but stifled through the revolt and fall of Adam, we became heir to "a hereditary corruption and depravity of our nature, *extending to all the parts of the soul*, which first makes us obnoxious to the wrath of God, and then produces in us works which in Scripture are termed works of the flesh." It follows "that being thus *perverted and corrupted in all the parts of our nature*, we are, merely on account of such corruption, deservedly condemned by God . . . " ([2], II, i, 8; italics added). It might be argued that Calvin, like Luther, was so profoundly impressed by Christ's atonement that he felt obliged to give account for its necessity as well as its benefit. Withal (and unlike Luther – who, although he also responded to problems, cast his principal theological doctrines in an affirmative mode) Calvin begins both sequentially and methodologically by affirming a problematic anthropology; and his task thereafter is to construct a theological theorem which is adequate to its solution.

God's sovereign will, wholly unconditioned and entirely determinative of all things, is Calvin's necessary remedy for this anterior diagnosis of the human condition; indeed, God's absolute sovereignty correlates with man's utter helplessness and is its functionally reciprocal relation. To say that 'depravity' is the *conditio sine qua non* of the human situation does not mean, according to Calvin, that the capacity to reason and will[5] has been entirely destroyed; it means, instead, that this capacity has been weakened, corrupted, "so enslaved by depraved lusts as to be incapable of one righteous desire" ([2], II, ii, 12). Thus bound and helpless and deservedly condemned, our liberty and righteousness and salvation are wrought through the atonement of Christ; by his perfect obedience, in life and in death, God's wrath is appeased and our redemption secured. In this gracious action "the Lord both corrects, or rather destroys, our depraved will, and also substitutes a good will for himself" ([2], II, iii, 7).

But not for everybody! Had this been a general dispensation, Calvin might have provided a warrant for beneficence; instead, the divine self-sacrifice is seen as efficacious only for those to whom God elects to give it. As a doctrine, predestination is uncomplicated and straightforward: "By predestination we mean the eternal decree of God, by which he determined with himself whatever he wished to happen with regard to every man. All are not created on equal terms, but some are preordained to eternal life, others to eternal damnation . . . each has been created for one or another of these ends . . ." ([2], III, xxi, 5). It thus appears that we have, not one, but two 'orders of creation'.

Knowledge of who is predestinated to what, of course, is God's secret; moreover, those preordained to eternal life differ in no visible way in this life from those preordained to eternal damnation. It would appear, therefore, that beneficent works are not only impossible but meaningless as well, and that human destiny is synonymous with fatalism. But the final chapter of the *Institutes* is "Of Civil Government"; and there Calvin tries valiantly to show a compatible and positive role for civil government, and thus for humanly directly beneficence. He acknowledges a distinction between civil government and 'heavenly kingdom' but declares that these two "are not adverse to each other." Indeed, Calvin is so bold as to state that "The former, in some measure begins the heavenly kingdom in us, even now upon earth, and in this mortal and evanescent life commences immortal and incorruptible blessedness" ([2], IV, xx, 2).

In so adverting to a natural theology which he had earlier repudiated, Calvin assigned the state responsibility for protecting life, guaranteeing a public form of Christian religion, and maintaining civil peace and order ([2], IV, xx, 2). Apart from asserting these things, however, he does not show how the role of government can be complementary to the Gospel; indeed, the greater weight of other arguments and proofs stands starkly against these declarations, and why they occur in this context is far from clear or certain. A possible bridge may lie in the notion that while good works cannot save, they may nevertheless be a sign that one has been 'predestinated to life'; but this was necessarily a very tentative hypothesis for Calvin, and a definite correlation between success and sanctity awaited the 'Yankee ethic' of American Puritanism. Another possible explanation may lie in Calvin's treatment of the authority of Scripture. He was not a literalist (as evident by his treatment of usury, polygamy, and holding goods in common), but he was a 'fundamentalist' for whom Biblical authors are "sure and authentic amanuenses of the Holy Spirit" ([2], IV, viii, 9), the "instruments" by which Scriptures come "from the very mouth of God" ([2], I, vii, 5). Since the elect alone are given power by the Holy Spirit to interpret Scripture aright, and since Calvin did not doubt that he was among the elect, it follows that his interpretations are authoritative! That might speak to the formal aspect of the matter; but why or how, in a material way, Calvin spanned the distance between devalued creation and salvation is not self-evident from his anthropological, soteriological, and theological predicates. The result, in sum, is more extravagant and repressive than with Luther because there is no secure place methodologically from which persons are either expected or permitted to exhibit charitable and kindly acts.[6]

It deserves saying, moreover, that any talk of good works as signs of election is methodologically very problematic inasmuch as there is no way, given native depravity and predestination as a secret of God, either to act beneficently or to identify which works can be called good. Calvin has not left any ground for that judgment or the discrimination it implies; and *if* consistent, he must be agnostic regarding what a good work is or even looks like. Max Weber probably captured the essence of the contradiction: " ... since Calvin viewed all pure feelings and emotions, no matter how exalted they might seem to be, with suspicion, faith had to be proved by its objective results in order to provide a firm foundation for the *certitudo salutis*. It must be a *fides efficax* ... " ([30], p. 114). If that be so, it is a small step from here to the doctrine that "God helps those who help themselves" and the concommitant notion that there is an obligation to succeed in business and commerce to the glory of God. Thus R. H. Tawney could observe that "It is not wholly fanciful to say that, on a narrower stage but with not less formidable weapons, Calvin did for the *bourgeoisie* of the sixteenth century what Marx did for the proletariat of the nineteenth, or that the doctrine of predestination satisfied the same hunger for an assurance that the forces of the universe are on the side of the elect as was to be assuaged in a different age by the theory of historical materialism" ([28], pp. 111–112). Calvin does, undeniably, provide a warrant for works; but, due to his theological diagnosis of illness and cure, he falls to provide similarly for beneficence.

III. WESLEY

Many extensions and permutations of the Lutheran and Calvinist reformations

– together, indeed, with a number of Protestant 'new beginnings' – occurred over the following two centuries; and among these, one of the more important contributions to a Protestant theology and ethics which warranted beneficence was John Wesley (1703–1791).

What Wesley seems to have learned at Aldersgate (24 May, 1738) was that justification is by grace alone and that this grace is given only through faith ([11], VI, pp. 7–8). He later defined faith as that which, given by grace, enables us to "feel [the] power of Christ . . . resting upon us" ([11], V, p. 167). More fully explicated, Christian faith is

> . . . not only an assent to the whole Gospel of Christ, but also a full reliance on the blood of Christ; a trust in the merits of his life, death and resurrection; a recumbency upon him as our atonement and our life, *as given for us, and living in us*; and in consequence hereof, a closing with him, and a cleaving to him, as our 'wisdom, righteousness, sanctification, and redemption,' or, in one word, our salvation ([11], V, p. 9).

Upon this predicate, one may safely say that Wesley's main concern (in contrast, for example, to that of Luther) is with the application and realization of the transforming power of saving grace within the life of the individual person. And while it may be true that Wesley takes his cue from the center of Luther's theology, namely, the doctrine of justification by grace, it may also be true that Wesley works out the full existential implications of Luther's crucial insight in a theologically and anthropologically more appropriate and balanced way.

But there is a certain danger in this kind of generalization. Although Wesley does certainly modify Luther's understanding of justification, one should not conclude thereby that Luther's paradoxical formulation of *pecca fortiter et crede fortius* or of *simul peccator, simul iustus* is irrelevant for Wesley's view. What is quite plain, however, is that Wesley wishes to distinguish clearly between the Christian life and justification. For Luther, all is accomplished with the act of justification; for Wesley, there is more. Charles Wesley's hymn puts the case succinctly:

> What shall I render to my God?
> For all His mercy's store?
> I'll take the gifts He hath bestowed
> and humbly ask for more ([16], No. 399).

John Wesley himself preached that

> . . . of yourselves cometh neither your faith nor your salvation: 'it is the gift of God;' the free, undeserved gift; the faith through which ye are saved, as well as the salvation,

which he of his own good pleasure, his mere favor, annexes thereto. That ye believe, is one instance of his grace: that, believing, ye are saved, another ([11], V, p. 13).

Thus, while grace is the source of salvation, faith is its condition; and God is the giver of the possibility of both.

With this in mind, it is appropriate to observe that Wesley's estimate of man, at least after his Aldersgate experience, is that there is nothing which man in himself alone can either achieve or perform which is deserving of the least thing from God's hand ([11], V, p. 7). Indeed, if man finds any favor at all with God, it is due entirely to God's own mercy. Hence, if man receives salvation, his only response can be "Thanks be unto God for his unspeakable gift" ([11], V, p. 8).

In describing Wesley's view of justification *sola gratia*, we have already spilled over into what is, at least implicitly, a statement of his understanding of the human condition. Perhaps one of the most lucid statements of the Wesleyan anthropology is found in another of Charles Wesley's hymns, "Jesu, Lover of My Soul;"

Just and holy is Thy name,
I am all unrighteousness,
False and full of sin I am,
Thou art full of truth and grace ([16], No. 110, v. 3).

The human predicament thus described is a universal reality inasmuch as "man was created looking directly to God, as his last end; but falling into sin, he fell off from God, and turned into himself" ([11], IX, p. 456). One is reminded here of Luther's notion of the *cor in curvatus in ad se*, producing a vicious circle of idolatries. Wesley used the same idiom.

Moreover, Wesley utilized, as had Luther, the Augustinian-federal theory of the Fall, and claimed that all persons are born into the state of separation by virtue of Adam's sin. Everyone inherits this corrupt nature; and, to this point, Wesley's thought strongly parallels that of both Luther and Calvin. "Man's heart," says Wesley, "is altogether corrupt and abominable" ([11], V, p. 7 and pp. 203–204).

Wesley does not claim that the truth of this assertion is universally obvious. In fact, *only* those whose eyes have been opened by God's grace can perceive their true condition. Grace, then, serves a double function in Wesley's theology: it is not only the means of establishing a right relationship with God; it is also finally the means for apprehending a proper self-understanding whereby one comes to recognize the *need* for a new relationship with God ([11], VI, p. 58).

One additional feature of Wesley's doctrine of man, in which he follows Luther and Calvin, should be noted: though fallen, man is nonetheless able to use his natural capacities for relatively good social purposes. These are not to be understood as ultimately good when judged by the normative standards of God's love and law. However, Wesley insisted upon a central point which, by way of emphasis at least, distinguishes his theological position from that of Luther and Calvin: in ethics as well as soteriology, concrete *relationship with God* (that of sinful alienation and/or faithful responsiveness) *is altogether determinative* for a theologically appropriate anthropology. In falling out of right relation with God, who alone is the ultimate power of righteousness, we inevitably fall into sin ([11], IX, pp. 456f). Even in the performance of relatively good acts, committed in relation to other men in society, our acts are performed out of a heart corrupted by wrong to our Creator.

Nevertheless, though in falling away from God we lose the 'moral image' of God, we by no means lose the image of God altogether. We do not lose this image altogether, *not* because we pull ourselves up by our own bootstraps, but because God, through prevenient grace, never lets us go. We remain in real relationship to God, even though this may be a sinfully perverted relationship. It is through the reality of this relation, as sustained by God himself, that the Creator continues, as it were, to inform us of the fact that we are 'made for' personal fellowship with our Maker and to offer us the responsible possibility of turning (or 'returning') to that fellowship ([11], V, p. 7). Thus we come to a more 'catholic' emphasis on prevenient grace which, in Wesley's understanding of its full existential implications, maintains the divine sovereignty while avoiding any necessity for postulating an ultimate predestination or divine determinism.

Although Wesley was in substantial agreement with Luther and Calvin at the point of insistence upon human moral impotency, and in ascribing our justification to God, *sola gratia*, he refused to agree that grace is particular and restricted in the ultimately selective and limiting sense implied by a predestinarian doctrine of election. To pose such a claim, according to Wesley, is to make God Himself unjust. In the place of such an election, Wesley understood the operation of grace to be "free for all and free in all" ([11], VII, p. 373 and pp. 380–381), bestowed in an appropriate way to each and every person, not merely to some restricted group whom God has foreordained.

Wesley perceived original sin as *man's* falling out of right relation with God; and he understood prevenient grace as *God's* continuing offer of the new *real possibility* of our responsive turning back toward that proper mode of relationship out of which we have fallen. As a result of his synthesis of these two doctrines, Wesley was able to hold together two seemingly disparate *foci*: man's inability to *move himself* toward God and his freedom to *respond* to God. Because of original sin, the natural man is "dead to God" and unable to move toward Him or to respond merely to the external proclamation of the Gospel; through the work of 'preventing grace,' however, he is given the power freely to respond either in resistance to or acceptance of God's mercy ([11], VII, p. 317).

Though Wesley insisted on *sola gratia*, he did not follow the psychology of the Wittenberg and Geneva Reformers to the point of claiming that there is *nothing* still operative within the life of the 'natural man' besides his natural qualities. In contrast to Luther and Calvin, Wesley claims that God endues man with 'preventing grace' which continually presents the human will with the *genuinely 'live' option* of responsively turning back again to God and receiving the saving grace of justification.

> ... No man living is entirely destitute of what is vulgarly called natural *conscience*. But this is not natural: It is more properly termed *preventing grace*. Every man has a greater or less measure of this, which waiteth not for the call of man. Everyone has, sooner or later, good desires; although the generality of men stifle them before they can strike deep root, or produce any considerable fruit. Everyone has some measure of that light ... which ... enlightens every man that cometh into the world ... so that no man sins because he has not grace, but *because he does not use the grace* which he hath ([11], VI, p. 512; italics mine).

Prevenient grace, then, is related to justification in quite the same manner as justification is related to sanctification; it is the necessary antecedent.[7]

Nevertheless, prevenient grace, although an evidence of God directly at work within even the natural man, is not enough to enable man to turn to God in the fullness of faith. God makes himself known to man in two quite different ways, in Wesley's thought: first, in a preliminary way, through conscience; and, secondly, through the proclamation of the Gospel. Hence, the work of salvation begins with the activity of prevenient grace upon the 'natural conscience.' And while Wesley acknowledges the third or instructing use of the law ([11], V, pp. 443–444), it would not be correct to say that in his judgment the law, *ex opere operato*, brings man to a recognition of his fallen state and thus to repentance. It is the role of prevenient grace, not the law, to lead man in his first positive step toward salvation. Thus, for Wesley, repentance or conviction of sin is existentially antecedent to faith.

I have argued that Wesley was in substantial agreement with both Luther and Calvin at the point of claiming that man in himself is morally impotent and that justification is by grace alone. I have attempted to show, further, that Wesley departed from both Wittenberg and Geneva in his claim that no 'natural' man *really* exists 'in himself' alone, that saving grace is free and unrestricted, and that freely responsive faith is required as the means of its appropriation. Now the other focus of Wesley's dialectic needs to be reasserted, namely, that salvation is the total work of God ([11], VI, p. 44). Regarding good works, Wesley insists that

> No works are good, which are not done as God hath willed and commanded them to be done; but no works done before justification are done as God hath willed and commanded them to be done: therefore, no works done before justification are good ([11], V, pp. 59–60).

If man is helpless, it is likewise true that he is void of righteousness and stands in condemnation before God. Still, through the merits of Christ, God has prepared for man's redemption. The foundation of justification, then, rests upon both Christ's righteousness and man's sinfulness ([11], V, pp. 54–56).

Being justified, however, does not mean being made actually righteous or just ([11], V, p. 56). It means precisely what it meant in the thought of the Reformers, namely, that we who have no righteousness are accounted as righteous.

> Justification ... is not the being made actually just and righteous. This is sanctification; which is, indeed, in some degree, the immediate fruit of justification, but, nevertheless, is a distinct gift of God, and of a totally different nature. The one [justification] implies, what God does for us through His Son; the other [sanctification] what he works *in us* by His Spirit ... in general use, they are sufficiently distinguished from each other ... ([11], V, p. 56; italics mine).

In more direct language, one might paraphrase Wesley to say that justification is the basis and beginning of the Christian life, that faith opens the way to the "new birth" and fullness of life in Christ, and that sanctification in holiness then follows.

In Luther and Calvin, both of these movements (i.e., repentance and trust in Christ) had been identified in justifying faith. Wesley, however, distinguished them, both ontologically and psychologically. The free response to prevenient grace indicates our readiness to receive God's further gifts of grace. The succeeding justifying faith is a sure trust and confidence in Christ which brings conviction of sin and forgiveness, and the awareness that we are saved by grace alone. While these two movements are distinguished in Wesley's thought, he strongly insists that they cannot be finally divided; that "at the same time that we are justified, yea, in that very moment, sanctification begins" ([11], V, p. 55). In other words, in the same moment that God makes His forgiveness known to the sinner, He begins the restoration to sonship. It should be plain that there is no quarrel between Wesley and the Reformers where the justifying act itself is concerned; for all three, it is God's act of pardon and acceptance ([11], V, p. 57). The difference occurs at the point at which Wesley insists that the act is *both* immediate and gradual, whereas interpreters of Luther and Calvin, while they have said both, have generally insisted that it must be *either/or*.

Wesley's understanding is that at the same moment that we are justified, we experience the new birth, which begins the process of sanctification, of being made righteous. Both a *relative* and a *real* change occur in the *same moment* ([11], V, pp. 233–234). Perhaps it would help to say that whereas for Luther justification connotes the whole content of salvation, for Wesley salvation is a process of which justification is a primary and basic stage.[8]

Wesley certainly approached Roman Catholicism from the point of view of this doctrine of 'double justification', where grace is not "given all at once ... but from moment to moment" ([11], V, p. 164). This was perhaps unavoidable when Wesley adopted the classical view of Christus Victor.[9] Yet, to the end of his writing, he guarded himself by insisting that there is no preparation for the entry of the Creator Spiritus; and Charles Wesley's hymn accurately reflects this essential tenet of his brother:

We cannot think a gracious thought,
We cannot feel a good desire,
Till Thou, who call'st a world from nought,
The power into our hearts inspire:
The promised Intercessor give,
And let us now Thyself receive ([16], No. 534, v. 2).

We must ask now, what is our part in all of this, what is the nature of human response, and how do men and women come to be repentant in Wesley's theology?

Faith is offered man as the free gift of a gracious God; but to this, man must actively respond by opening himself to the possibility of receiving that gift. We cannot say, then, that in Wesley's conception grace is either solely an apportionment by God or merely an appropriation by man. Indeed, in the last analysis, it is both. To God alone is ascribed the power and the glory manifested in man's justification; but to man alone is reserved (by God!) the right of decision, made possible by prevenient grace, to accept or to reject God's offer. So, Wesley argues in the 1770 *Minutes*:

> We have received it as a maxim, that 'a man is to do nothing in order to justification.' Nothing can be more false. Whoever desires to find favor with God should 'cease from evil, and learn to do well' . . . ([15], p. 96).

Obviously contrary to Luther and Calvin, Wesley stressed that repentance and works "meet for repentance" are the precondition of justifying grace. But he guarded himself from both Antinomianism and Pelagianism.[10]

> God does undoubtedly command us both to repent, and to bring forth fruits meet for repentance; which if we willingly neglect, we cannot reasonably expect to be justified at all; therefore both repentance, and fruits meet for repentance, are, in some sense, necessary to justification. But they are not necessary in the same sense with faith, nor in the *same* degree; for those fruits are only necessary *conditionally*; if there be time and opportunity for them. Otherwise a man may be justified without them . . . but he cannot be justified without faith; this is impossible . . . repentance and its fruits are only *remotely* necessary; *necessary in order to faith*; whereas faith is *immediately* and *directly* necessary to justification ([11], VI, p. 48).

The nature of human response, then, is simply to reach out and claim the gift of faith; but always with the acknowledgment that grace is the enabling source of this freely active response. Wesley's reasoning was that if God is working in you, as indeed He is, there is no excuse for your not working, too; indeed, it is imperative that you do so ([11], VI, p. 509 and pp. 512–513). He did not attempt to explain precisely how this comes about, except to say that God provides some power by which one is enlightened and may accordingly see his true state, acknowledge God's judgment, and confess his utter helplessness ([11], V, pp. 102–103). But this enlightenment is sufficient to bring man to repentance; and this is the first step in the process of justification. The second follows in the form of positive belief in the truth of the gospel. Wesley rejects, however, any notion of mere "partnership" with God and insists upon the continually initiating and empowering activity of God's grace which makes man's genuinely free response (in faith or unfaith) a real possibility.[11]

IV. BENEFICENCE IN AMERICAN PROTESTANTISM

I have so far suggested that a viable concept of beneficence is not readily available from the theologies of Luther and Calvin because, despite admonitions to do good works to the glory of God out of gratitude, both consistently ignore that the Christian life is marked by works of love as the fruits of faith. Wesley, on the other hand, clearly understood and regularly insisted that faith itself is inauthentic if not witnessed to and nourished by charitable acts.[12] I have proposed, moreover, that the possibility for works of charity is theologically warranted when it is appreciated – as Wesley did, but Luther and Calvin did not – that justification includes liberation from *both* the guilt *and* the power of sin; and moreover that works of love keep faith alive.

An assessment of (especially American) Protestant ethics and beneficence cannot end here, however, if only because we are left with no accounting for how it is that American Protestants are so heavily invested and engaged in actions which are intended to serve the social, political, and economic well-being of the neighbor. To understand why this is so, we must comment briefly on certain important modifications in American Protestantism of the theologies of Luther and Calvin; and that, I think, is most helpfully done by acknowledging the ways in which the eighteenth century Enlightenment affected American Protestant theology and ethics. In this aspect, I want to suggest that American Protestantism is the offspring of an alliance between the sixteenth century Continental Reformation and the eighteenth century English Enlightenment. In so saying, there is no need to suppose that either Luther or Calvin would (or could) have anticipated the triumph of rationalism and empiricism in English-speaking Protestantism; nor is there any need to trace *all* of the particularities (and peculiarities) of American Protestantism to Luther and/or Calvin. The facts weigh heavily on the side of American Protestantism having become what it is through an extended and intimate liaison between the Continental Reformers and the English Enlightenment Philosophers.

Liberal Traditions in Colonial Theology

While 'rationalism' and 'empiricism' have often been ranged against each other, they were not so adversarial in the eighteenth century. Instead they allied themselves to each other in the common task of setting human intelligence against divine revelation, human rights against entrenched aristocratic privileges, and progressive social change against custom and tradition. St. Augustine had held that mankind was originally perfect, but then suffered a fall from grace which rendered Adam's progeny both incapable of improving its state and wholly dependent upon God's Providence to redeem us out of the earthly into the heavenly city. Some centuries later, St. Thomas had affirmed a high place for reason but, similarly owing to the effects of original sin, did not hold it to be the sole criterion for truth. The authority of revelation, he held, was *reasonable*, but not *reasoned* human intelligence; and revelation, in his view, was thus clearly superior to reason.

But in the eighteenth century, philosophers held reason to be the judge of all things, including the value of Church and Bible; and the marks of the 'reasonable Christianity' of the Enlightenment developed accordingly. (1) Typical of the Enlightenment is the idea that the order of nature is a coherent structure, pervaded throughout by natural law. *Design in nature*, which inductively leads to Nature's God, and which places emphasis on objective order rather than subjective consciousness, is a predominating motif. Far from diminishing humanistic tendencies, confidence in man's rational capacity (to discover God through Nature's design) actually reinforced them. (2) Revealed religion is not eliminated in this scheme. Newton and others stressed that *natural religion is the foundation* which relevation, while it may go beyond, cannot contradict. In the 'radical' wing, it was insisted that nature alone is the medium of revelation; in the 'moderate' wing it was maintained that revelation was needed to 'ratify' or 'supplement'. Withal, both agreed that, via natural religion, we can know that there is a God, that he demands virtue, and that rewards and punishments are part of the scheme of things. (3) If it is the right of the mind to investigate, and if reason is the tool by which nature's meaning and intention are discovered, it follows that inquiry must be free and that a variety of opinions must be tolerated. The act of 1689 is thus not only the result of developments toward constitutional monarchy but toward the acceptance of *religious toleration and pluralism* as well.[13] Through the philosophy of Locke, it came to be held not only that variety of opinion is normative but that the Church is also built upon the principle of voluntarism. America inherited these notions; they became cultural mainstays, and their influence upon contemporary American Protestantism is now nowhere disputed. (4) As a rational, self-conscious, self-determining agent, *man has dignity and worth* as that being who is the determiner of what determines him. Though he joins with others and forms societies, he does so principally in order to secure individual self-realization. The human self is highly regarded because he is guided from within and not coerced from without. To the eighteenth century rationalist, the doctrine of the fall and original sin was an offensive depreciation of man, because inherited sin lodged squarely against the rationalist denial of organic dependency and affirmation of individual responsibility. (5) Finally, there is a movement through all of the preceding hallmarks toward *doctrinal simplicity*. Not only were traditional doctrines unimportant and treated peripherally but, in their place, preeminence was given to character and the good life, so that 'reasonable Christianity' consisted chiefly of ethically directed love of God and love of neighbor. This step, derivative from but in concert with the others, efficiently completed the separation of Christian theology *qua se* from ethics, and thereby legitimated an autonomous moral philosophy which, while it did not cavil at Christian principles, found it unnecessary to appeal to them either for authentication or for authority.

When did these ideas become elements in the thinking of Americans? It is certain that, as early as the latter part of the seventeenth century, Americans were becoming concerned with natural or 'experimental' philosophy. Specifically, we know that in the late seventeenth century John Winthrop, Jr., (who was later made a member of the Royal Society) brought from England a telescope which was used by Thomas Brattle (then treasurer of Harvard) in 1680 to observe a comet. Brattle's findings in this observation were communicated to Newton, who later announced the theory that comets' courses are determined by gravity. Meanwhile, to Cotton and Increase Mather, this comet of 1680 was related to some foreboding event; and while it would later be explained in terms of natural phenomena, to the Puritans it was the manifestation of some religious and moral significance. There was, of course, no distinction among these Puritans between primary and secondary causation; but the scientists around Harvard at the turn of the eighteenth century increasingly viewed natural phenomena in terms of natural philosophy and science. The role of such events as mysterious portents was giving way, and the supernatural was yielding to the natural: natural law best explains natural events.

As it happened, 1680 was an important year in colonial America for yet another reason: it was in that year that John Wise (1652–1725) was called to begin a new church in Chebacco parish (Essex County, Massachusetts), where he remained until his death. Well-known for his fierce independence, Wise was jailed, fined, and suspended from the ministry for his protest against a tax levied by Governor Edmund Andros; and when, in 1705, Cotton and Increase Mather were instrumental in securing the endorsement of several of the Massachusetts ministerial associations for a paper called *Proposals,*[14] Wise reacted vigorously to what he perceived to be the curtailment of traditional New England parish autonomy. In 1713 Wise published his satirical response, *The Churches Quarrel Espoused*, and in 1717 a sequel, *A Vindication of the Government of New-England Churches*. The purpose of both documents was to show that Congregationalism of the ancient sort is the *true* Congregationalism; and that the 'proposals' of the Mathers, *et al.*, would destroy local church autonomy.

Part II of Wise's *Vindication* is particularly interesting, for our purposes, because it is so thoroughly permeated with the philosophy of the Enlightenment: man is considered in "a state of natural being . . . a free-born subject . . . owing homage to none but God himself . . . properly the subject of the law of nature . . . [for whom] reason is congenate with his nature . . . [possessing a] great immunity [which] is an original liberty instampt upon his rational nature . . . [which further guarantees] an equality amongst men . . . " ([31], pp. 21–48). So prominent, indeed, is the Enlightenment philosophy in this document that scholars of this period have observed that "The unique feature of this notable treatise lies in the fact that it defends the 'New England Way' by appeal to 'the Law of Nature'. Wise may therefore well be called the "Morning Light of the American Age of Reason" ([26], I, p. 384).

The significance of John Wise is not exhausted, however, by noting his advocacy of 'reasonable Christianity' in the American colonies. As importantly, he anticipated the American Revolutionary policy of majority rule and, additionally, related the religious principle of local church autonomy to the civil government, thereby widening the breach between church and state.

Meantime, several distinguished proponents and opponents joined this debate, the fundamental issue of which lay in the interpretation of the religious affections versus the rational understanding. George Whitfield, Gilbert Tennent, Samuel Hopkins, *et al.*, ably represented the 'enthusiast' branch of the Great Awakening; while Jonathan Mayhew, Charles Chauncy – and eventually Jonathan Edwards – insisted upon a reasonable and intelligible accounting of all acts of affection as somehow also constituting acts of will.

Chauncy and Edwards, for example, contended that the warrants for their theologies were not derived from the principles of Enlightenment philosophy but from a heavy dependence upon serious biblical study. Nevertheless, Edwards *was* an advocate of double predestination, while Chauncy thought that doctrine unreasonable and inconsistent! At issue in their disagreement was, of course, more than scriptural exegesis; the real stakes were set by their hermeneutics.[15]

In his *The Benevolence of Deity*, Chauncy challenged the Edwardian theology to answer how a morally good God could order and produce eternal misery for some of his creatures, and to show how eternal misery for the creature manifests moral goodness in the Deity. Chauncy argued that benevolence in God must be "exerted under the direction of intelligence and in consistency with fit conduct" ([4], p. 33). Injustice, falsehood, and deceit (all immoral acts, he said) cannot possibly be interpreted as being fit to promote the good of Creation; indeed, not even God, by a sovereign act of will, can constitute good as evil or evil as good ([4], p. 33). "Benevolence in the Deity," said Chauncy, "signifies precisely the ... disposition freely to communicate all the good that is consistent with wise and fit conduct" ([4], p. 39). Man – endowed as he is with common sense (or moral discernment), self-determination, and conscience – is "efficiently the cause of his own volitions and actions" ([4], p. 130), and, by implication, able to judge the reasonableness of God's benevolence. Thus it is not only true (as Edwards held) that we are free to do as we choose, but also (as Edwards denied) that we are free to choose what we will! "The exertions of our minds and bodies are under our own dominion" ([4], p. 136); we are the cause, and the proper cause, of our own volitions.

That is a long way, of course, from predestination and foreordination; and Chauncy, in many respects wanting to conserve and extend New England Calvinism, tellingly displays how effectively the die had already been cast against old-line, orthodox Calvinism. By the turn of the nineteenth century, a considerable group of Enlightenment liberals (strong enough to wrest from the orthodox both the Hollis Chair of Divinity and the Presidency of Harvard) could be found in Boston and its environs.

The extent to which religious advocacy of Enlightenment liberalism influenced the American revolution is beyond our interests here, but there need be little doubt that both sacred and secular interests in Colonial America found their unity in an emerging quest for freedom – the former from oppressive and stultifying orthodox theologies, and the latter from tyrannical and burdensome monarchy. Indeed, orthodox and rationalist Calvinists frequently found themselves united in defense of independence and liberty; a solidarity which was eventually to overcome otherwise important doctrinal formulations in fealty to the ideal of the free individual, who enjoyed a God-given immunity from interference as he pursued his own rational interests. On this point, the religious and political communities were practically indistinguishable; both envisioned that, whatever else, America's chief ends included the extension and perfection of individual liberty. That religious orthodoxy should be a casualty of this conflict of ideologies seems retrospectively to have been unavoidable. In any event, by 1820 – the year following William Ellery Channing's sermon, "Unitarian Christianity," which was the *magna carta* of the liberal movement in nineteenth century America – colonial (i.e., Calvinistic) religion had been replaced by an indigenous American religion of 'secular Christianity'.

V. PROTESTANT ETHICS, BENEFICENCE, AND HEALTH CARE

To find morality detached from religion is nothing new in the history of Western civilization; and to find this circumstance in American Protestantism is not a novel disclosure, although other treatments of this phenomenon (known to me) have generally dealt with the matter thematically.[16] Though in a somewhat adumbrative brevity, I have wanted to suggest how the lineaments of classical Continental Protestant theology, when recast according to the philosophy of the English Enlightenment, understandably issue in liberal and reasonable religion. Indeed, it would be a serious error, I think, to suppose that the agenda of the advocates of 'reasonable Christianity' ever extended to the abolition of religion; they wanted, instead, merely to put religion in its 'proper place', and that was a place amenable to the tenets of the Enlightenment.

It is not always self-evident how moral maxims gain credibility and authority in an entire society, and thereby become the foundations upon which both personal and institutional relationships rest. But in the nineteenth century, rationalism wed to experimental scientific methodology provided answers to that query in America. Assuming that religious faith no longer fitted human experience, it represented a relatively small step (although surely taken in view of impressive evidence) to suppose that scientific rationality could provide the kinds of authority and certainty which, when extended to man's moral being as well as to the natural universe, would merit belief in *it* as the only way to truth. Owen Chadwick's summary of the result of secularization in Europe is no less true in America:

> Moral principles could exist, did exist without religion; even when they were the moral principles of the religion. The complicating fact for the late nineteenth century was the claim that you could have morality without Christianity while the morality which you must have was Christian morality. But men were good Christian men morally without possessing Christian faith ([3], p. 237).

Subsequent movements in American Protestantism – punctuated by the struggle over slavery, two world wars, the Social Gospel, recurrent revivals of one sort or another, and more recently by electronic evangelism and liberation theology – have tended to confirm and extend liberal traditions, even when hypostasized in fundamentalism or other 'back to the Bible' crusades. The operational hallmarks of American Protestantism continue to be individualism and autonomy, and their necessary social corollary, pluralism. With conspicuous exceptions – I think, for example, of Walter Rauschenbusch's version of the Social Gospel (*not* to be confused with that of the likes of Washington Gladden and Richard T. Ely), or of Reinhold Niebuhr's 'Christian realism' in the early 1940's, both of which refused reduction of Christian witness to American nationalism – the gospel of American Protestantism has generally been quite congenial to American social, political, and economic interests. This has been so prominently the case, in fact, that religion in America appears to possess little capacity for transcending the national consciousness, for singing the Lord's song in a strange land. Perhaps it is precisely the familiarity between religion and culture over more than two-and-a-half centuries that now breeds contempt for any religious expression which is not homogeneous with the distinctively American ethos. Indeed, it was precisely that recognition that led Reinhold Niebuhr to observe that America is 'at one and the same time, one of the most religious and most secular of nations" ([21], pp. 138–139).

In the 1950's and 1960's, Protestant seminarians were exposed to this criticism from a variety of sources, prominent among which was Will Herberg's *Protestant-Catholic-Jew*; but that and similar texts are now either largely unknown or ignored because the thesis is so existentially self-evident. All the same, after a quarter-century, Herberg's diagnosis of American civil religion still rings true:

Americanness today entails religious identification as Protestant, Catholic, or Jew. . . . To be a Protestant, Catholic, or a Jew are today the alternative ways of being an American. . . . Yet it is only too evident that the religiousness characteristic of America today is very often a religiousness without religion, a religiousness with almost any kind of content or none, a way of sociability or 'belonging' rather than a way of orienting life to God ([9], pp. 274, 276).

I graduated from seminary in the year that Herberg's book was published; and there may be a touch of irony in the fact that my class was simultaneously exposed to another volume, this one written by a bishop in one of American Protestantism's prominent denominations, who declared:

Our church is suited to the American temperament, which it has always expressed and embodied in all its actions and viewpoints. . . . The big . . . faults [of our church] – and we have them, let us confess – are the big American faults; and the big . . . virtues [of our church] – and thank God, we have them too – are the big American virtues. . . . When Theodore Roosevelt was president, he once privately remarked to the chaplain of the United States Senate, who was [of our church], 'Your church is the church of America' ([8], pp. 21–22).

So the revolutionary spirit of the eighteenth century continues to play an important role in the history of American Protestantism, and American Protestantism continues to reinforce an indigenous American consciousness. Robert E. Fitch once wrote that "There is a Protestant strength, and its name is liberty. There is a Protestant sickness, and its name is anarchy" ([6], pp. 498–505). Perhaps this essay is no more than an extended commentary on Dean Fitch's observation and indictment.

There is sufficient evidence that the American social and political ethos has significantly influenced American Protestantism; it is also the case that American Protestantism has affected the professions and the social and political institutions of this land. How might this bear particularly upon American health care?

Paul Tillich once said that *the Protestant principle* "is the guardian against the attempts of the finite and conditioned to usurp the place of the unconditional in thinking and acting" ([29], p. 163). He meant by this the rejection of 'heteronomy' (the attempt of a religion to dominate autonomous cultural creativity from the outside) and the repudiation of 'autonomy' (whereby a civilization's ties with its ultimate ground and aim are cut and secular humanism is affirmed) and, in their place, the avowal of 'theonomy' (which he defined as a culture in which the ultimate meaning of existence shines through all finite forms of thought and action) ([29], p. xii). The Protestant principle, in other words, is a protest against any absolute claim made for a relative reality, any ideology which serves the 'man-made-god' of social group, class, or nation.

In practice, Tillich's notion of the Protestant principle has had an altogether negative service: when the ultimate meaning is luminous in finite forms of thought and action, it must be rejected as heteronomous and/or autonomous, because an absolute claim is corrupted when communicated by (or in or through) a relative reality.[17] Among its most eminent theologians,[18] the Protestant principle functions as *protest* (i.e., to object or to disapprove) rather than *protestari* (i.e., in the Latin, to declare or witness in public); but that is a necessary emphasis when human experience is bereft of any ultimate ground inasmuch as that ultimate ground is only approached when those human experiences are shattered by *angst* and its corollaries. It may sound harsh to say, but on these terms the Protestant principle is misanthropic, and beneficence presents chiefly as expiation from human depravity.

American Protestantism has so effectively divorced theology from ethics that no distinctively authorizing ground for moral decision-making is typically offered, apart from or beyond the cultural ethos. Denominational statements on current problems in medical care (chief among which are various aspects of human sexuality, including abortion and homosexuality) are therefore typically *ad hoc* and loosely – if at all – contextualized in a body of ecclesial doctrine. Correspondingly, Protestant ethics *qua se* is able to offer little or no advice to physicians and others in the health care sector which is, in principle, based on other than the prevailing values resident in the liberal ethos of a pluralist and secular society. Because 'relative reality' has been sundered from 'ultimate meaning', there is little alternative to offering counsels of despair when issues are critical, and councils of accommodation when issues are academic.[19]

Philosophers, lawyers, physicians, and social scientists have only recently addressed themselves to moral issues arising from health care; and Protestant religious ethicists were only somewhat earlier in their inquiries into these matters. Roman Catholic moral theologians, on the other hand, have long been attentive to medico-moral problems, and orthodox Judaism possesses a long tradition of Rabbinic responsa and the dicta of religious courts. The first significant book to appear in the United States in the twentieth century, by other than a Roman Catholic or Jewish author, was Joseph Fletcher's *Morals and Medicine* in 1954 [7]. Sixteen years later, Paul Ramsey's *The Patient as Person* appeared [23], as did my own book, *Ethics and the New Medicine* [25]. In the succeeding decade we have witnessed a veritable deluge of articles and monographs and symposia from a variety of intellectual and professional interests; and prominent among these have been contributions from religious ethicists. The large majority of these scholars, I think it safe to say, are alumni of graduate religious studies programs in schools and universities long associated with one or another of America's Protestant churches. Consistent with the Protestant principle, it is relatively rare in my acquaintance with this literature that authors appeal specifically to either theological or ecclesial authority as warrant for asserting moral claims; alternatively, those warrants are sometimes cited as interesting historical landmarks. These essays, instead, typically appeal to the liberal traditions of American Protestantism for both analytic and normative axiological concepts, with the result that, once again, the value assumptions and judgments offered to physicians and other medical professionals are chiefly those of conventional cultural wisdom. Given the professional disposition of many physicians to ethical autonomy, there is little reason to expect that they want more; given the atheological disposition of many religious ethicists, there is little reason to expect that they will be able to offer more.

Large and urgent matters, freighted with ethical significance, already confront us in the health care sector; and the future promises more, not less, moral quandary. Abortion, recombinant DNA research and application, large scale clinical investigations, delivery of primary care, national health insurance, allocation of scarce high technologies, redefinition of professional and paraprofessional roles – these and many other issues are currently upon us, competing for attention and advocates and advantage. Each and all of them are argued by some of us to do us good; and each and all of them are argued by others to do us ill.

But no less large and no less urgent than these matters is the question upon which their answer depends: where will we look to find out what words like good, bad, right, wrong, charity, enmity, beneficence, and maleficence mean? And what will warrant and authorize the meanings we want to assert and claim?

Our liberal heritage directs us away from univocal definitions and toward pluralism, away from communitarian and toward individualistic meanings, away from notions of common good and toward propensities calculated to achieve my own good. Political consensus in our society is thus achieved by requiring only that each individual be guaranteed protection to pursue his or her own interest fairly; if there is a 'good' to be sought and served, it is neither more nor less than this. In the absence of any shared, common conception of the good which is larger than freedom from interference to pursue one's own interests, it is impossible to judge the morality of conduct (acts) because we have already abdicated any judgment of character (agents). Among the systemic reasons which account for this disposition are two which merit explicit mention: the liberal tradition has demonstrated an unwillingness or inability (or both) to name *the good* toward which human moral agency in social, political, and other efforts ought to move; and lacking a well-defined *telos*, the capacity for freedom has been typically understood as an end rather than a means.

Our social experience, however, increasingly confirms that the virtues of liberalism – autonomy, individualism, pluralism, secularism – whether exercised by individuals or professional organizations or 'corporate persons', are a centrifugal force, calculated *not* to draw us together but to pull us apart. Such a force, we can agree, may have produced expansion in times of unlimited frontiers; but now, in a time defined by limits to growth, it produces adversity and cleavage. Our problem, in a word, is that we lack a rational method for resolving moral conflict in liberal society, because we cannot or will not agree among ourselves on the goal(s) of our moral agency. Moreover, autonomy, individualism, pluralism, and secularism are inimical to a *common* good and unable to be translated into public policies which will function centripetally. Our mistake is not our valuing of freedom, but our understanding of it.

Some among us have ventured to describe our planet as 'spaceship Earth', or to speak of our common human condition within a 'global village'. To use these metaphors is to talk of commonality, concord, covenant – categories of reference that are richly endowed in the Jewish and Christian religious traditions of the West, but currently somewhat out-of-style. I believe that none of the urgent (and presently awesome) health care matters before us is insoluble if we can agree upon the predicates and purposes of health care. But it is a profound problem for us just now, rooted in the 'reasonable Christianity' of the Enlightenment and reinforced by the cultic piety of American Protestantism: we cannot, as Ramsey so tersely put it, agree upon the practice of virtue because we will not agree upon the principles of virtue.

Duke University
Durham, North Carolina

NOTES

[1] I want to acknowledge with genuine thanks assistance received during the preparation of this essay to Ms. Susan L. Pate, the Rev. Dr. R. Taylor Scott, and Ms. Bettye W. Smith.

[2] This is not an observation which is unique to medicine; indeed, for a variety of reasons, the development of modern professionalization in virtually all of its modes has tended toward isolation, insularity, and self-sufficiency. For the standard diagnosis of this phenomenon, cf. [27]: "In our society (that is, advanced western society) we have lost even the pretence of a common culture. Persons educated with the greatest intensity we know can no longer communicate with each other on the plane of their major intellectual concern. This is serious for our creative, intellectual and, above all, our moral life. It is leading us to interpret the past wrongly, to misjudge the present and to deny our hopes of the future. It is making it difficult or impossible for us to take good action" ([27], p. 59).

[3] Cf. esp. Luther's "Address to the Christian Nobility of Germany" in [13].

[4] Cf. Luther's "A Treatise on Good Works" in [13].

[5] Calvin's psychology supposes that reason and will, *ratio* and *arbitrium*, are correspondent and bilateral moral functions in a person: "It is a power of reason to discern between good and evil; of will, to choose the one or other" ([2], p. 227).

[6] H. Richard Niebuhr has done much to popularize the idea that Calvin belongs – together with Augustine, Wesley, F. D. Maurice, and others – to the 'conversionist' or 'transformationist' type in the Christ-culture debate. Even granting that typologies necessarily generalize, Calvin fits very loosely (if at all) into this one. Niebuhr argues that conversionists share certain theological convictions about creation, fall, and history which affirm the opposition between Christ and all human institutions and customs but do not therefore "lead either to Christian separation from the world ... or to mere endurance in the expectation of a transhistorical salvation ... " Instead "Christ is seen as the converter of man in his culture and society;" and Calvin is portrayed as one who believed "that what the gospel promises and makes possible, as divine (not human) possibility, is the transformation of mankind *in all its nature and culture* into a kingdom of God in which the laws of the kingdom have been written upon the inward parts" (cf. [20], pp. 42 and 190ff. Italics added). Such a view, as I have suggested, is not supported by the data.

[7] Cf. the entry in Wesley's *Journal* for September 13, 1739 ([11], I, pp. 224–225).

[8] Cf. [12], p. 84f.

[9] See, for example, Wesley's sermon on "The End of Christ's Coming" ([11], VI, pp. 367–377). The text is I John 3: 8. "For this purpose was the Son of God manifested, that he might destroy the works of the devil."

[10] Antinomianism is the view that, for Christians, the requirements of the law have been abolished. Literally 'against law', antinomianism holds that obedience to the law has been rendered unnecessary since one is justified, not by works of the law, but by grace alone through faith in Jesus Christ. 'Pelagianism' does not reject the concept of grace, but it "asserted that man as accountable must retain some freedom and power of action towards moral growth and that the function of grace, therefore, is educative and co-operative." Cf. [14], pp. 14, 140.

[11] Portions of this section on Wesley have been excerpted from [24], pp. 120–128.

[12] Cf. Wesley's condensation of the Church of England's *Certain Sermons or Homilies, Appointed to Be Read in Churches*, in ([22], pp. 130–133). Authorized by Edward VI for publication in 1547, these *Homilies* represented the Anglican consensus regarding the doctrine of salvation, faith, and good works.

[13] The Toleration Act of 1689 was the first statutory grant of religious toleration in England, and ended the Church of England's monopoly of the nation's religious life. It did not increase the number of Dissenters, but it did permanently weaken the authority of the Church. Thereafter, religious pluralism became a fact of English life. This act followed the displacement of Roman Catholic James from the throne, and the accession of Protestants William and Mary. Cf. [10], pp. 17–49 and [17], pp. 100–123.

[14] This document included, for example, provision that associations could pass upon the qualifications of ministerial candidates, or examine cases of suspected scandal or heresy.

[15] Rudolf Bultmann has shown the importance of acknowledging preunderstandings and predispositions which are brought to Biblical interpretation in a brilliant essay, "Is Exegesis Without Presuppositions Possible?" He answers his own question with an emphatic "no!" Cf. [1].

[16] Cf., e.g., [18] and [19]. Both books attempt essentially the same program, *viz.*, to interpret the movement and meaning of American Christianity; but the earlier one abjures an approach via theological doctrines in favor of history, sociology, and ethics,

whereas the latter one tries specifically to take account of the force of Christian doctrine which produced the diversity in American religion.

[17] It may be worth noting, in passing, that precisely this understanding accounts for so much eucharistic theology which fails to appreciate that it is we, not simply (or even importantly) the elements, who are transfigured and transubstantiated in Holy Communion.

[18] I think, for example, that in addition to Tillich the same tendency is evident in the work of Karl Barth, Reinhold Niebuhr, and others who insist upon radical discontinuity between divine and human. Sometimes this is expressed as an 'infinite qualitative distinction', *totaliter aliter*, as in Barth; sometimes it takes the form of a thoroughgoing disjunction between 'sacrificial' and 'mutual' love, as in Niebuhr.

[19] It is not my task here to offer an alternative theological ethics which might overcome these deficiencies, although I acknowledge that this is an item which deserves to be prominent on the agenda of Christians in America. On the other hand, these observations pertaining to the American and Continental Protestant and liberal religious values should be noted as part of a much wider historical and cultural tide of criticism, especially as regards the philosophical assumptions and beliefs associated with political theory and practice. Radical critiques with which I find myself in congenial company (without necessarily agreeing with their solutions) have been published by Robert Paul Wolff, Hannah Arendt, Sheldon Wolin, Michael Oakeshott, Isaiah Berlin, Michel Foucault, among others.

BIBLIOGRAPHY

1. Bultmann, R.: 1957, 'Ist voraussetzungslose Exegese möglich?', *Theologische Zeitschrift* **13**, 409–417.
2. Calvin, J.: 1957, *Institutes of the Christian Religion*, transl. by H. Beveridge, James Clarke and Co., Ltd., London.
3. Chadwick, O.: 1975, *The Secularization of the European Mind in the Nineteenth Century*, Cambridge University Press, Cambridge.
4. Chauncy, C.: 1784, *The Benevolence of Deity* (Tract), Boston.
5. Erasmus, D.: 1945, *Essai sur le Arbitre*, transl. of *De libero Arbitrio* by P. Mesnard, Les éditions Robert et Pevé Chaix, Algiers.
6. Fitch, R. E.: 1966, 'The Protestant Sickness', *Religion in Life* **35** (Autumn), 498–505.
7. Fletcher, J.: 1954, *Morals and Medicine*, Princeton University Press, Princeton.
8. Harmon, N. B.: 1955, *Understanding The Methodist Church*, The Methodist Publishing House, Nashville.
9. Herberg, W.: 1955, *Protestant-Catholic-Jew*, Doubleday, Garden City.
10. Horwitz, H.: 1977, *Parliament, Policy, and Politics in the Reign of William III*, Manchester University Press, Manchester.
11. Jackson, T. (ed.): 1829, *Wesley's Works*, 3rd edition, John Mason, London.
12. Lindstrom, H.: 1946, *Wesley and Sanctification*, The Epworth Press, London.
13. Luther, M.: 1883ff, *D. Martin Luthers Werke. Kritische Gesammtausgabe*, H. Bohlaus, Weimar.
14. Macquarrie, J. (ed.): 1967, *Dictionary of Christian Ethics*, Westminster Press, Philadelphia.

15. Methodist Conference Office: 1812, *Minutes of The Methodist Conferences, from the first, held in London, by the late Rev. John Wesley, A. M.*, London.
16. Methodist Conference Office: 1933, *The Methodist Hymn-Book*, London.
17. Miller, J.: 1974, *The Life and Times of William and Mary*, Weidenfeld and Nicholson, London.
18. Niebuhr, H. R.: 1929, *The Social Sources of Denominationalism*, The Shoe String Press, Hamden, Connecticut.
19. Niebuhr, H. R.: 1937, *The Kingdom of God in America*, Harper and Bros., New York.
20. Niebuhr. H. R.: 1951, *Christ and Culture*, Harper, New York.
21. Niebuhr, R.: 1952, 'Prayer and Politics', *Christianity and Crisis* **12** (October 27), 138–139.
22. Outler, A. C. (ed.): 1964, *John Wesley*, Oxford University Press, New York.
23. Ramsey, P.: 1970, *The Patient as Person*, Yale University Press, New Haven.
24. Smith, H.: 1964, 'Wesley's Doctrine of Justification: Beginning and Process', *The London Quarterly and Holborn Review* (April), 120–128.
25. Smith, H. L.: 1970, *Ethics and The New Medicine*, Abingdon Press, Nashville.
26. Smith, H. S., *et al.*: 1960, *American Christianity*, Charles Scribner's Sons, New York.
27. Snow, C. P.: 1963, *The Two Cultures*, Cambridge University Press, Cambridge.
28. Tawney, R. H.: 1926, *Religion and the Rise of Capitalism*, John Murray, London.
29. Tillich, P.: 1948, *The Protestant Era*, The University of Chicago Press, Chicago.
30. Weber, M.: 1930, *The Protestant Ethic and the Spirit of Capitalism*, transl. by T. Parsons, George Allen and Unwin Ltd., London.
31. Wise, J.: 1772, *A Vindication of the Government of New England Churches* (Tract), Boston.

SECTION III

BENEFICENCE IN HEALTH CARE

NATALIE ABRAMS

SCOPE OF BENEFICENCE IN HEALTH CARE

This paper considers some of the implications of positing a duty of beneficence in the health care context, with particular emphasis on the scope or boundaries of such a duty. Part I discusses alternative sources or groundings for the duty of beneficence. Part II presents various definitions of beneficence and considers them in relation to health care. Part III focuses more directly on the central issue, the general scope or limitations of the duty of beneficence in the health care context.

The assumption is made, at least in part, that beneficence should be considered a duty of the agent(s) in question, rather than beyond the call of duty, i.e., supererogatory. It would seem that this assumption is implicit in the title for this essay. If there is a scope or boundary to beneficence in the health care context, then beneficence cannot be seen as completely supererogatory since this would imply a limit to what in principle can or should be allowed to be done to assist another. Surely we would not want to place an a priori limitation on supererogatory acts of beneficence. Talk of the scope of beneficence therefore seems to include the assumption that at least some acts of beneficence are part of one's duty. The difficult task is to define the boundary between beneficent acts that are required by duty and beneficent acts that go beyond what is required.

I. SOURCE OF BENEFICENCE AS A DUTY

A significant question concerns the possible source(s) for such a duty. At least three different bases for a duty to be beneficent can be identified ([1], p. 141).

First, the duty could actually be based on a prior duty of reciprocity. That is, individuals might be thought to have the duty of beneficence toward others because of benefits which they have already received. Such an argument is frequently given with regard to participation in biomedical research. Here it is argued that because we are all beneficiaries of medical discoveries made possible by the contributions of other individuals, we all have obligations to participate as subjects in some form of research. This basis has been seen to be so strong that it has even been applied to the use of children and

Earl E. Shelp (ed.), Beneficence and Health Care, 183–198.

fetuses in research [9]. Because of the numerous benefits which children and fetuses receive as the result of prior research, it has been argued that it is acceptable to use them as research subjects, despite their inability to offer informed consent. Two arguments which can be made against the reciprocity claim are suggested by Hans Jonas [5]. First, if the original participation or performance of beneficent acts was voluntary, then the appropriate response would be gratitude, not obligation. One does not have a duty to repay a voluntary act of kindness, although it might be especially nice to do so. Thus, with regard to research, for example, a duty to participate would not follow from prior voluntary sacrifices. Second, if a duty of repayment were required, the repayment would be owed to those individuals who made the original sacrifices. Again, with regard to research, participation would not be appropriate, since those who would benefit would be future generations and not the individuals who made the original sacrifices. Similarly, with regard to the usual therapeutic relationship between physician and patient, it is rarely the case that the physician owes a debt of gratitude to a given patient.

Another basis which has been suggested as the source of duties in the health care context is the needs of others. According to this theory, a physician, for example, has a duty of beneficence which emanates from the patient's needs. However, as pointed out by Beauchamp and Childress ([1], p. 142), the needs of individuals are not usually considered sufficient for the imposition of either the legal or the moral duty of care. It is recognized both by the law and in the AMA *Principles of Medical Ethics* (1957 and 1980 Codes) that a physician can choose whom he or she will serve. The existence of any need per se does not automatically impose a duty of care. It is only subsequent to the establishment of a doctor-patient relationship that the duty of care is applicable. Exceptions to this principle are emergency situations and, at least from a moral point of view, some situations in which a need can be satisfied with a minimal amount of effort or risk on the part of the physician.

The question of whether needs ever constitute the grounds for a duty to render health care relates more generally both to the problem of good Samaritanism and to the issue of the distribution of health care. Usually the stringency of the moral duty to be a good Samaritan depends upon several factors: what the resultant harm(s) will be if help is not rendered; what the costs or risks of rendering care will be for the agent; how imminent the harm(s) will be if help is not received; and how likely it is that an option open to the agent will actually alleviate the situation. It is most often argued that if the harm to be alleviated is severe, the risk(s) or cost to the agent minimal, the harm(s) imminent, and the likelihood of an improved outcome great, an individual has at least the moral duty to render aid. With regard to the doctor-patient relationship, this moral duty becomes a legal one, only when a professional relationship has been established, at which time omissions become legally punishable.

Several arguments are given [6] for not making the rendering of aid in certain circumstances (i.e., acting beneficently) a legal responsibility. It should be noted that in most European countries rendering such aid is legally mandated. One argument presented against what is called good-Samaritan legislation is that it would be much too difficult to draft from a legal point of view, in that it would be difficult for such legislation to guard against a demand for unreasonable sacrifices. Another fear seems to be that such legislation would constitute an encroachment on freedom and privacy and would be counter to the emphasis on the individual and individual rights, so highly valued in American society. Further reasons cited against the enactment of good Samaritan legislation involve the difficulty in determining whether an individual guilty of not rendering aid actually knew of the other's problematic situation, the fear that such legislation amounts to an enforcement of morality which can only be done to prevent the actual infliction of harm, and the belief that legislating morality will inevitably alter individual motivation and force such behavior out of fear of the law rather than from a purely altruistic desire to aid another. Whatever the merits of these arguments may be with regard to legally mandating good Samaritanism, such behavior seems required from a moral point of view, at least given the conditions previously mentioned. To use Judith Thomson's words, whereas a physician need not be a 'good Samaritan', he/she should at least be a 'minimally decent Samaritan' [13].

The role of needs as the basis of a duty to render care, or act beneficently, is addressed from a different point of view by Bernard Williams. With regard to the distribution of health care, Williams claims that "it is a matter of logic that particular sorts of needs constitute a reason for receiving particular sorts of goods" ([14], p. 129). When the need which should be the basis for a just distribution of a good is not operative, i.e., when the need itself is not sufficient to secure the good, there exists an irrational state of affairs. Since, for Williams, the proper ground of the distribution of medical care is ill health (he claims this is a 'necessary truth'), when illness is not sufficient to secure care, but wealth is also necessary, the situation is irrational, not merely immoral or wrong, in the sense that in such a situation the appropriate reason (ill-health) is insufficiently operative. According to Williams, it is a situation insufficiently controlled by reasons.

Concerning Williams's position, it is important to draw a distinction between what constitutes basic needs versus felt needs or wants. Surely individuals have interests in and desires for various goods which we would not want to call needs. To say that a person has a need for something is to claim that the good is essential to the person's welfare, i.e., to keep the individual at a minimum level of well-being. As such, medical care, in most instances, would be legitimately classified as a need, although certain cases might be excluded, e.g., some instances of cosmetic surgery and orthodontia, and perhaps even some instances of infertility treatment. Clearly delineating what kind of health care should be considered necessary as the result of human need and what kind should be considered beneficial but not necessary would be an extremely difficult task. Furthermore, it is basically because the need for health care is so extensive that distribution problems have arisen. As suggested by Outka [10], Williams's argument that the proper ground of the distribution of medical care is ill health must be modified in view of the scarcity of resources. Rather, illness should be seen as the proper ground for the 'receipt' of health care. Its distribution in cases of scarcity forces us to select another basis for allocation.

A third possible source for duties in the health care context is the particular contracts or promises which are made, either explicitly or implicitly. The question then becomes whether the duty of beneficence follows from the so-called 'contract' or promises which are made in establishing the doctor-patient relationship, or from a general social contract which can be presumed to exist between all individuals in society.

Do physicians have a duty of beneficence toward patients as a result of their particular roles and the professional relationship or as a result of their duty to act beneficently to all individuals, as joint members of the same society? Do individual patients have a right to be treated beneficently as the result of their particular position of hopelessness in the doctor-patient relationship or as a result of their general rights as human beings? Although it is beyond the scope of this paper to fully answer these questions, the direction of an answer can be suggested. If by acting beneficently physicians are expected to render aid which goes beyond what non-physicians could or should provide it would seem as if their duty to so act can be imposed only as a result of their particular roles. Requiring beneficence as a duty from health care professionals does not follow from a putative duty of beneficence in ordinary relationships.

If beneficence is to be seen as a duty in the health care context, its source is thus best seen as the implicit or explicit agreement between the practitioner and the individual patient. Neither a prior duty of reciprocity nor the needs per se of individual patients (except perhaps in circumscribed situations) provides an adequate basis for such a duty.

II. DEFINITION OF BENEFICENCE AS A DUTY

Assuming that there is a duty of beneficence in the health care context, the next question concerns the definition of beneficent behavior. Following traditional normative theory, beneficent conduct could be defined in at least two different ways: according to the agent's intentions or the consequences of the act. If the agent's intentions are definitive, an act would be beneficent if the agent intends to help an individual, regardless of whether the individual in fact benefits. If the consequences of an act are definitive, an act would be beneficent only if the individual in question benefits, whether or not the good outcome was intended by the agent. For present purposes, the question then becomes how the duty of beneficence should be defined in the health care situation.

It seems clear that a patient's improved health status, if not intended by acts of the health professional, cannot be viewed as the result of the agent's beneficence. If no effort is made to improve the patient's condition, the health professional cannot be praised for beneficent conduct. (Intentional harm will not even be considered in this relationship). Similarly, good intentions, in and of themselves, although necessary to the definition of beneficent conduct, do not appear sufficient. It is also necessary for the physician's good intentions to result in actual good for the patient in order to view the physician's conduct as beneficent. There are at least two different interpretations of intention, however, when it is said that a patient's improved health status is intended by the physician. On the one hand, the patient's improved condition could be the sole motivating factor or intention of the physician. Alternatively, it could be intended by the physician but only as a means toward some other ulterior goal, whether it be money, prestige, professional advancement, etc. Therefore, even if intention is thought to be a necessary component of beneficent behavior, there is still the question of how this intention functions as a motivating factor.

Joel Feinberg discusses this distinction ([2], p. 296) and identifies three ways in which a person can have an interest in the well-being of another. In one case, an individual can be interested in the well-being or improvement of a second person because the latter's status indirectly relates to or affects the condition of the former. It is in the agent's interest that the interests of the other individual be advanced. In a second case, an individual has an interest in the well-being or improvement of a second person directly. The agent has such a strong personal investment in the other person's condition that the latter's interest actually becomes an ulterior interest or 'focal aim' of the agent himself. In this second situation, the agent's interest in promoting the individual's well-being is not simply as a means to promoting the agent's own condition, but is an end in itself.

Feinberg points out the difference between this second situation and a third case, which he calls the "common phenomena of spontaneous sympathy or pity" ([2], p. 291). In this common situation of 'genuinely disinterested compassionate motives', which can be directed at total strangers, although an individual agent may act to try to relieve the unfortunate circumstances of another person and although the suffering of the other individual may make the agent extremely unhappy, the harm that has been done to the suffering individual is not also harm done to the agent. To the extent that the agent suffers, his feelings are vicarious. In the second situation described above, however, the agent has an intrinsic interest in the well-being of the other individual, and the latter's misfortune directly affects the agent's well-being. Improving the unfortunate individual's situation is an end in itself for the agent.

It is quite difficult to decide whether or not these different types of situations are instances of beneficence. Case one, however, seems the least problematic. If a physician acts to improve a patient's condition simply in order to upgrade indirectly his own condition (either by obtaining payment, recognition, or whatever), the physician's behavior does not appear to be a clear instance of beneficence. On the other hand, if improving the physician's situation is an added benefit or side-effect of helping the patient, but is not the sole motivating factor, then it is more reasonable to view the physician's actions as beneficent. According to some, the second case may appear to be the best instance of beneficence. Here the physician's desire to aid the patient would be an end itself. However, this situation has come about only because the patient's interests are so intimately connected to those of the physician himself or herself. It may be argued, therefore, that such a situation is probably not one of true beneficence since the goal can be seen to be improving the agent's own condition, not that of another. A similar controversy exists over whether parental behavior in the interests of children should be seen as beneficent. To the extent that improvement in the child's welfare is felt by the parents to be a direct benefit to them, it is questionable whether such behavior should be viewed as beneficent.

Case three seems to present the clearest instance of beneficence. In this situation, a physician acts with the sole intention of improving the patient's condition, which can deeply disturb the physician, and yet the patient's illness has not also become a harm to the physician himself. The harm which the patient suffers is not internalized by the physician. Furthermore, in this case, the agent's actions are not motivated by a desire to obtain indirect rewards. It is for these reasons that case three appears to be the best instance of beneficent behavior. This type of interaction or involvement with another person's well-being would also seem to be the most conducive to the rendering of optimum care. In this type of relationship, although the physician cares directly about the patient's condition and intervenes to improve it, the so-called 'objective' stance, which is thought to be so important in the physician-patient interaction, is still maintained. If the physician were to become overly involved in a patient's condition, as might be the case if the patient were a family member, it is possible that the physician might lose the ability to evaluate objectively the patient's condition. Moreover, if this happened regularly, the physician could become emotionally exhausted and not be able to care properly for his or her other patients. An act which falls into this third category would seem to coincide with the definition of kindness suggested by Marvin Kohl. According to Kohl, an act would be kind (the term 'beneficent' is being considered here as synonomous with 'kind') because it: (a) "is intended to be helpful; (b) is done so that, if there be any expectation of receiving remuneration (or the like) the individual would nonetheless act, even if it becomes apparent that there is little chance of his expectation being realized; and (c) results in beneficial treatment for the intended recipient" ([7], p. 91).

III. SCOPE OF DUTY OF BENEFICENCE

So far it has been suggested that there is a duty of beneficence in the health care context and that the concept of beneficence is best defined in accordance with Kohl's definition of kindness and with Feinberg's third case. What has not been discussed, however, are the limits, if any, which should be placed on the duty of beneficence in this context.

Several attempts have been made to define the scope of the duty of beneficence. One such attempt is made by Peter Singer [11]. Singer argues that individuals are obliged to act beneficently unless this would mean that something of 'comparable moral importance' would have to be sacrificed. In other words, if it is in an individual's power to prevent something bad from happening, and he or she can do this without producing further harm (at least of comparable weight), then the individual has a duty to act. Applying such a broad scope to the duty of beneficence of health care professionals would be extremely demanding. It would imply that a health professional is always under an obligation to render aid, as long as the sacrifice which the agent would have to make does not exceed the harm which would be prevented. Since the health professional's potential sacrifice would rarely equal, and even less likely ever surpass, the degree of harm or illness to be alleviated, the health professional would always be obligated to come to the aid of all sick individuals.

Recognizing the excessive demands of such a broad limitation, Michael Slote [12] proposes a much narrower principle. According to Slote, an individual is obliged to prevent harm only when it is possible to do so without seriously interfering with one's life style, provided that one's life style does not itself involve acts which are morally wrong. Therefore, one could morally justify not preventing harm in those situations in which providing assistance would disrupt one's life, even if the harm which would be prevented is of considerably more weight, e.g., death, than the sacrifice which would have to be made. Applying such a limitation to the health care context still imposes enormous obligations. Rarely is it necessary for health professionals seriously to disrupt their life styles in order to render aid. Of course, this is very much a matter of degree. Although helping one individual in a given situation might not interfere in any way with one's life style, rendering assistance whenever it is needed might very well cause considerable disruption. A significant problem with Slote's principle is its failure to consider and make a recommendation with regard to the cumulative effect of such beneficent actions. Where does one's duty end? At what point can it be said that a series of relatively small beneficent acts interferes with the agent's lifestyle.

A third principle of beneficence is proposed by Tom Beauchamp and James Childress. They state that "X has a duty of beneficence toward Y only if each of the following conditions is satisfied: (1) Y is at risk of significant loss or damage; (2) X's action is directly relevant to the prevention of this loss or damage; (3) X's action would probably prevent it; and (4) the benefit that Y will gain outweighs any harms that X is likely to suffer and does not present more than minimal risk to X" ([1], p. 140). One of the primary differences between Singer's and Slote's principles and this one is that whereas Singer and Slote each define the scope of beneficence by comparing the harm to be alleviated with the degree of required sacrifice, Beauchamp and Childress's principle attempts to define these two factors independently.

They claim that the risk to the agent must not be more than minimal, and the harm to be alleviated must be significant loss or damage. Despite the remaining difficulty of defining 'minimal risk' and 'significant loss or damage', at least on this third principle, the health professional would not be obliged to render aid simply because the harm to be prevented was large relative to the risk involved. On this view, even if the differential between the harm to be prevented and the sacrifice is great, the health professional would not be obligated to act unless the risk to the agent were in fact minimal and the risk to the patient was considerable. Here it is not simply the differential that matters.

An additional factor to consider as a possible limit on the duty of beneficence, one that is ignored by all of the above principles, is the patient's or recipient's wishes. If the person to be helped does not want to receive assistance, does this actually negate the duty to render aid or does it create a practical obstacle which prevents the agent from performing what still remains to be his or her duty? The answer to this question depends upon the primacy accorded the duty of beneficence in the doctor-patient relationship.

On one interpretation, the duty of beneficence could be based upon, or be derivative from, a patient's right to be treated in a certain way. According to this view, the patient's right to beneficent treatment is primary, and the physician's duty is secondary to or follows from this right. Those who view the duty of beneficence in this way generally see patient autonomy as an actual limit to such a duty. A second interpetation is that, conversely, the physician's duty to act beneficently is primary, and it is because of this duty that a patient has the right to such treatment. On this view, patient autonomy is generally seen as a practical limitation which prevents health professionals from doing their duty. A third interpretation is that the duty of beneficence follows neither from a primary duty of the physician nor a primary right of the patient. Rather, such a duty exists because recognizing it produces the best consequences, essentially a utilitarian argument. Health professionals are required to act beneficently because their doing so is the best way to encourage the optimum production and distribution of health care. Whereas in most situations, the particular rationale for a duty of beneficence may not be significant, it becomes important in situations of conflict. For example, if in a given instance honoring a patient's wishes would not be protecting the patient, and if acting beneficently is primarily based on a physician's duty, there would be a conflict between this duty and a patient's interest in autonomous decision-making. Similarly, there could be a conflict between producing the best consequences, if that is brought about by acting beneficently, and a patient's right not to be treated paternalistically.

Different weighting or ranking of the principles of beneficence versus autonomy can have other far-reaching practical implications. Physicians are frequently faced with the question of how far to pursue a patient who either refuses the recommended medical treatment or agrees in principle but is nevertheless noncompliant. If the duty of beneficence is considered primary, even if the physician believes the patient has given informed consent, he or she may be required to go to considerable lengths to try to convince a patient to accept or follow treatment. If the principle of autonomy is considered primary, and again it is believed that the patient has made an informed choice, the physician would not be obligated by the duty of beneficence to pursue the matter any further. The patient's wishes would curtail or limit his or her duty of beneficence.

In addition to the patient's refusal, which has been said to be either a practical hindrance to fulfilling the duty of beneficence or a limitation on the principle per se, there are at least two other factors which can limit in varying ways the health professional's duty to act beneficently.

One other limitation on the duty to be beneficent is ignorance. Since beneficence requires that good consequences actually result, ignorance of what will produce positive results places limits on a physician's ability to act beneficently. If the ignorance is shared by the profession, i.e., if it is simply a lack of available scientific or medical knowledge, then the ignorance is obviously excusable. If, however, failure to improve a patient's well-being is a function of the individual physician's ignorance of readily available information, the ignorance does not serve as an excuse for not fulfilling one's duty of beneficence.

Another type of ignorance can also result in impairing a patient's well-being. A physician must also have knowledge about the personal values and preferences of the individual patient. It is not sufficient for the physician to know simply the severity or magnitude of alternative procedures. Deliberations about whether a particular benefit/harm ratio is acceptable necessarily includes two steps. First, the nature of the harms and benefits, their severity or magnitude, their duration, and probability of occurrence must be known. Second, assignment must be made of the values or utilities associated with the harms and benefits. Given the pluralistic nature of our society, it is not reasonable to assume that everyone places the same value on potential harms and benefits, and their probabilities. It is therefore necessary for the physician to be aware of the patient's value system. Ignorance along these lines could be equally as harmful as ignorance of medical science.

Two additional related factors which could limit a physician's ability to fulfill his or her duty of beneficence is a lack of necessary resources and /or the competing rights and interests of patients other than the one of immediate concern (either other patients of the physician in question, or patients of other physicians, or any physician's potential patients). Obviously, even with the best of intentions, a physician cannot act beneficently if he or she does not have the means to do so. This lack of resources may be the result of numerous factors, including insufficient physician time, personnel, hospital space, machinery, and/or medications. With regard to the problem of the allocation of resources and the issue of beneficence, the main question is whether the duty of beneficence requires a physician to render optimum personal care to his or her individual patients or whether the duty obligates the physician to render as much aid as possible, even if this means compromising, to some extent, the treatment rendered to any one patient. Essentially the conflict is between the principles of distributive justice and beneficence, and the question is whether the principle of distributive justice limits or defines the scope of either allowable or required beneficence.

An excellent article addressing this question with regard to the legal profession is 'The Lawyer as Friend; Moral Foundations of the Lawyer-Client Relation" by Charles Fried [4]. Fried addresses essentially the same issue with regard to medicine in parts of his book, *Medical Experimentation, Social Policy and Personal Integrity* [3]. The argument which Fried basically supports is that all professionals, whether physician, nurse, lawyer, or clergyman, are required by such professional norms as the duty of beneficence to act in the exclusive best interest of the patient or client. However, some professionals argue that such an approach renders the relationship at times immoral and at best amoral. Two reasons are generally cited. First, such an approach constitutes a social harm because it wastes scarce resources, professional time, and effort. Exclusive attention to the best interests of individual clients and patients (i.e., greatly limiting the beneficiaries of one's beneficence) fails to confer benefits wisely, efficiently, and most importantly, equitably. As Fried states, the charge is generally that "the ideal of professional loyalty to one's clients/patients permits, even demands, an allocation of the professional's time, passion, and resources in ways that are not always maximally conducive to the greatest good of the greatest number. The professional ideal authorizes a care for the client and the patient which exceeds what the efficient distribution of a scarce social resource (the professional's time) would dictate" ([4], p. 1061).

A second reason offered against this traditional model of professional loyalty is normally seen to apply primarily to the legal professional. It claims that such an approach constitutes not only social harm, but also harm against identifiable or specific individuals. By arguing exclusively for one's client without regard to truth or justice (e.g., by attempting to discredit the truthful testimony of a reliable witness) one is thereby directly harming another, the adversary. Although an adversarial relationship per se does not seem to exist in medical care, there definitely are instances in which exclusive attention to a patient's best interests produces harmful consequences for identifiable others. When a physician works to maintain the life of a severely deformed infant this frequently is at the emotional, psychological, and financial expense of others. Other instances include situations in which a second party is harmed because of a physician's pledge to keep a patient's information confidential. An interesting case which highlights this problem is one in which two individuals who are engaged to marry each other are both the patients of a single physician. One of the patients is homosexual, but does not want the other to know this fact. The physician has to weigh the patient's right to confidentiality against what he/she considers to be in the interest of the other patient [8]. Another case poses the adversarial or conflict of interest situation slightly differently. In this case, a husband returns from a business trip abroad, calls his wife's gynecologist of fifteen years to tell the physician that while he, the husband, was away, he contracted gonorrhea from a one night fling. The husband wants the gynecologist to protect the wife from any possible harmful physical consequences without telling her what he is doing. In all these instances, acting in the exclusive best interest of one party could produce harmful consequences for another. Such conflict situations are therefore not exclusively within the province of the legal profession. Whereas the first criticism accuses the professional of failing to benefit unidentified individuals, this second criticism accuses the professional of harming an identified adversary or other.

A second general question presented by the tradition of narrowly defining the scope of a professional's duty of beneficence and limiting professional responsibility to individual care is whether such an approach places any restrictions on the kinds of clients or patients a professional should accept. Again two arguments can be made in this regard. First, if a professional is going to devote exclusive attention to the best interest of a limited number of patients or clients, is the professional under any kind of obligation to choose patients or clients from among those who have the greatest need? Should the professional have almost complete discretion, as is presently the case, concerning what cases to accept or reject or does the duty of beneficence mandate certain behavior in this area? As noted previously, codes of professional responsibility, as well as the law, clearly reserve this right for the professional, stipulating only that a professional must act in the patient /client's best interests. Second, given this basic professional responsibility, not only is it permissible, but might it be obligatory for a professional to reject certain individuals as patients or clients. For example, should a lawyer defend a client whom he/she knows to be guilty, or a client whose goal is clearly one which the lawyer cannot morally support, as might be the case in a situation in which a lawyer is asked to work for the legal rights of a business to establish a nuclear power plant. Similarly, should a physician ever refuse to care for an individual who has been found guilty of a serious offense? Does the duty of beneficence demand that a physician render health care to any one in need, regardless of individual merit or worth?

A third problem is raised by viewing the duty of beneficence as mandating a professional relationship based upon exclusive personal care or attention. If the goal of a professional is to act with the sole intent of benefitting the patient/client, does this imply that the professional is morally justified in using or obligated to use all the legal means available to achieve this goal? With regard to the legal profession, this is an extremely serious issue. Examples abound. Should a lawyer allow a client or a witness who will commit perjury to take the stand when it is probably true that if the person does not testify the jury will decide against the client? Should a lawyer encourage an innocent client to plead guilty in order to receive a reduced penalty? Should a lawyer cross-examine a witness for the purposes of discrediting the reliability of a witness who is known to be telling the truth?

In medicine, the question of immoral yet not illegal means is less significant, but examples do exist. To what extent should a physician, acting in accordance with the norm of the patient's best interest, falsely document the existence of illness in order to have an individual found exempt from the draft or some other responsibility? Alternatively, should a physician ever fail to record a certain health problem (e.g., epilepsy or cancer) if having such information on a health record would be detrimental to a patient's employment prospects? To what extreme should a physician go in securing scarce resources for his or her patient, e.g., by trying to use status or rank in order to obtain the last available dialysis machine or place in the I.C.U.? Confidentiality is again an important issue here. To what extent should a physician maintain confidentiality in the patient's interest when another individual might be harmed, as in the newly recognized obligation of the psychiatrist to warn an individual about the patient's potential for harming that person, or when the public might be harmed, as in the case of an airline pilot suffering from a heart condition who is threatened with losing his job if his health status is revealed?

Charles Fried makes an interesting and worthwhile distinction between wrongs which are personally inflicted and those which are inflicted by the institutional system. Lying to a judge, humiliating an adverse witness, and using status to obtain a place in the I.C.U. would all be instances of personally inflicted harms. Not revealing confidential information, using the statute of limitations or a tax loophole would be instances of wrongs inflicted by the institutional system itself. Although the professional may not approve of the system's rules, Fried claims it would constitute negligent conduct for a professional to refuse to use the system to maximize the benefit for the patient or client. The professional would be absolved from wrong in such situations because the client or patient himself could have utilized the system in such a way were it not for lack of technical information or expertise. According to Fried, refusing to aid patients or clients in utilizing the system to the greatest extent possible would only work to discriminate against the less educated. Personal harms, however, are entirely different, and although the borderline is certainly not clear, the professional should not be seen as obligated to use his own person, as opposed to his professional capacity, in any questionable way in order to benefit the patient or client.

Two general arguments can be made in favor of the tradition of the professional-patient relationship, i.e., one that limits the scope of the duty of beneficence to the individual patient or client. One argument, albeit not a moral one, in favor of the traditional stance is that such an approach is psychologically easier in that it eliminates many, although not all, moral conflicts. If one has a single clear goal or end point, moral conflicts are not as perplexing or agonizing. Limiting one's accountability as much as possible limits conflict. The plight of a nurse, employed by a hospital, vividly demonstrates the agony of multiple accountability. The nurse is accountable as a human being to her/his own conscience, as a citizen to the state, as an employee to the hospital, as an assistant to the physician, as a professional to his or her own professional guild or standards, and as a professional to the patient.

The second argument in favor of the traditional view, and the one that seems the most convincing, is that limiting the duty of beneficence to specific individuals is the best method to recognize the rights, autonomy, and integrity of the individual. The traditional roles of medicine and law are designed in order to help re-establish, in the best possible way, the personal autonomy which is shattered by both illness and crime or the denial of legal rights. It is not simply that exclusive attention to the individual actually contributes in the long run to the collective good, i.e., it is not simply a utilitarian argument. Rather, exclusive attention to the individual is warranted by a desire each one of us has, to be recognized as valuable, responsible, and free moral beings. Precluding exceptional circumstances, on this view the scope of the duty of beneficence in health care mandates that the physician-patient relationship and the concept of personal care should be held inviolate.

IV. CONCLUSION

Although the relationship which has been the focus of this paper and which most readily comes to mind in discussing beneficence and health care is that between the physician and patient, it should be recognized that this relationship is only one among many in which the duty of beneficence is relevant. When considering the role of beneficence in health care, one must also consider the relationships among different health professionals, between society and health professionals, society and the individual patient, individual patients and each other, and society and future generations. Furthermore, each one of these relationships must be considered as reciprocal, i.e., it is important to consider not only what role beneficence has to play in the physician's (or any professional's) behavior toward an individual patient but also what role it has to play in the patient's behavior toward physicians. Similarly, in considering the relationship between an individual and society, the question of the role of beneficence must be applied both ways. In exploring the role of beneficence in these other relationships, one would have to include consideration of such things as an individual's duty to donate blood or organs, to contribute financially to medical research, and to volunteer as subjects in research, a health professional's duty to provide some care free of charge, society's duty to provide home care to the elderly and disabled, and society's duty to conserve and preserve resources and to try to achieve advances in health care for future generations. Unfortunately, it is not possible to go beyond a very few aspects of the physician-patient relationship in this paper and to suggest some general problems in defining the scope of beneficence. Hopefully, future work on this topic will include these additional relationships which define the health care context.

New York University School of Medicine
New York, New York

BIBLIOGRAPHY

1. Beauchamp, T. and Childress, J.: 1979, *Principles of Biomedical Ethics*, Oxford University Press, New York.
2. Feinberg, J.: 1977, 'Harm and Self-Interest', in P. M. S. Hacker and J. Raz (eds.), *Law, Morality and Society*, Clarendon Press, Oxford, pp. 285–308.
3. Fried, C.: 1974, *Medical Experimentation, Social Policy and Personal Integrity*, American Elsevier, New York.
4. Fried, C.: 1976, 'The Lawyer as Friend: Moral Foundations of the Lawyer-Client Relation', *The Yale Law Journal* 85, 1060–1089.
5. Jonas, H.: 1969, 'Philosophical Reflections on Experimenting With Human Subjects', *Daedalus* 98 (Spring), 219–247.
6. Kleinig, J.: 1976, 'Good Samaritanism', *Philosophy and Public Affairs* 5 (Summer), 382–407.
7. Kohl, M.: 1974, *The Morality of Killing*, Humanities Press, New York.
8. Kuschner, H.: 1977, 'Case Study: The Homosexual Husband and Physician Confidentiality', *Hastings Center Report* 7 (April), 15–17.
9. McCormick, R.: 1974, 'Proxy Consent in the Experimentation Situation', *Perspectives in Biology and Medicine* 18 (Autumn), 2–20.
10. Outka, G.: 1974, 'Social Justice and Equal Access to Health Care', *The Journal of Religious Ethics* 2, 11–31.
11. Singer, P.: 1972, 'Famine, Affluence, and Morality', *Philosophy and Public Affairs* 1, 229–243.
12. Slote, M.: 1977, 'The Morality of Wealth', in W. Aiken and H. La Follette (eds.), *World Hunger and Moral Obligations*, Prentice-Hall, Englewood Cliffs, N.J., pp. 124–141.
13. Thomson, J.J.: 1971, 'A Defense of Abortion', *Philosophy and Public Affairs* 1, 47–66.
14. Williams, B.: 1971, 'The Idea of Equality', in H. A. Bedau (ed.), *Justice and Equality*, Prentice-Hall, Englewood Cliffs, N.J., pp. 116–137.

EARL E. SHELP

TO BENEFIT AND RESPECT PERSONS: A CHALLENGE FOR BENEFICENCE IN HEALTH CARE

I. INTRODUCTION

The belief that health care should be a beneficent enterprise is explicit in the "Principles of Medical Ethics" adopted by the American Medical Association in 1957. The first section states that "the principal objective of the medical profession is to render service to humanity with full respect of the dignity of man" ([1], p. 39). It is not unreasonable to suggest that this objective would be acceptable to almost all health care personnel and institutions. The ideal and norm of 'service tempered by respect' is one worthy of moral admiration. If service is accurately interpreted to mean benefit, then a grounding of this objective in the principle of beneficence seems warranted.

Benefiting and respecting persons are noble goals for any human activity. They seem particularly appropriate to health care which aims, among other things, to contribute in countless ways to human well-being. People who provide care and cure to others are frequently characterized as benevolent and their ministries as beneficent. These ascriptions seem appropriate in most cases even though the full meaning of beneficence in health care contexts may be incompletely understood or expressed. As is true of other moral principles, beneficence is not subject only to one definition or interpretation. The confusion which may attend an application of the principle to health care settings may be due, in part, to certain conceptual uncertainties. Further, efforts to do the beneficent thing may pose practical problems of moral significance. One can reasonably question if the imperatives to 'benefit' and to 'respect' do not at times counsel contradictory means and ends of human conduct.

Sally Gadow illustrated how interpretations of moral principles applied to clinical settings seem to conflict in such a way that the requirements of both cannot be fulfilled at once. She contrasted the principle of beneficence with the principle of autonomy and concluded that autonomy has priority over beneficence in situations of conflict. She stated that " ... at times, autonomous patient decisions defy medical definitions of benefit, ... In those cases it is necessary to recognize that distinct principles are in conflict and a choice must be made, both cannot be satisfied at once" ([12], p. 683).

Earl E. Shelp (ed.), Beneficence and Health Care, 199–222.

This perception of the either/or nature of the options mandated by the principle of beneficence in situations in which other moral principles may be relevant, especially autonomy or respect for persons, seems to rest on an incomplete understanding of the principle of beneficence.

The primary purpose of this essay is to examine the meaning of beneficence and its application to health care. Particular attention is given to interpretations of benefit and respect. The application is limited to the patient-physician form of the therapeutic relationship. The meaning of the principle of beneficence for institutional arrangements and health policy is the subject of other essays in this volume. This is a conceptual inquiry which has implications for practical dilemmas. Indeed, understandings of conceptual requirements influence practice, and perceptions of realities and possibilities influence conceptual understandings. One area cannot be explored profitably without a vision toward the other. In order to begin to understand beneficence, I shall explicate first the principle of beneficence which will entail a consideration of some of the variables and qualities or conditions that are ingredient to its interpretation and which also contribute to the confusions associated with it. Second, I shall apply the findings of the conceptual survey to the context of health care and comment on the application.

It is necessary to note several assumptions and limitations before beginning. First, I assume that all moral agents have duties of beneficence. Duties of beneficence are more identified in some specific roles, such as health care, that some individuals voluntarily assume. Thus, I do not consider the limits of beneficence or the distinction between beneficence and supererogation. Both concerns are discussed in other essays in this volume. Second, my discussion assumes situations involving mature, competent, rational individuals without defining these important qualifiers. The special considerations of beneficence for the fetus, children, and others (e.g., comatose, mentally incompetent, mentally retarded) are specifically excluded. Third, I do not discuss the relationship of beneficence to paternalism. I prefer to view paternalism, at best, as a form of beneficence which requires a justification apart from an appeal to beneficence. The two may share certain characteristics but are not interchangeable. This survey is limited to the meaning of the principle of beneficence and its application to health care.

II. PRINCIPLE OF BENEFICENCE

The principle of beneficence commonly is understood to require one to do good and to prevent harm. The principle is distinguished from benevolence which refers to one's disposition or character trait to will the good or to delight in the good of another. William Frankena understood beneficence to be composed of four prima facie duties that he listed in lexical order, other things being equal: (a) one ought not to inflict evil or harm, (b) one ought to prevent evil or harm, (c) one ought to remove evil, and (d) one ought to do or promote good ([12], p. 47). I agree with several others, including Tom Beauchamp and James Childress ([2], p. 135), that the proscription against inflicting evil or harm is better understood as a separate, more stringent principle of non-maleficence. The prevention of evil or harm, the removal of evil, and the doing or promoting of good then become prescriptions or principles of positive beneficence. A segregation of imperatives along these lines is instructive to an understanding of beneficence in health care. This positive thrust is seemingly reflected in the statement of the American Medical Association with which the essay began.

Beauchamp and Childress revise the usual statement of the principle in such a fashion as to direct attention to critical elements of beneficence that could be overlooked otherwise. They define beneficence as "the duty to help others further their important and legitimate interests when we can do so with minimal risk to ourselves" ([2], p. 136). They place two principles under a general heading of beneficence. One is a principle of positive beneficence which requires the provision of benefits. The second is a principle of utility which requires a balancing of benefits and harms. Further, they see the principle of utility negotiating between beneficence and non-maleficence in situations in which they conflict ([2], pp. 136, 144). It is important to note several phrases in their definition that are central to its meaning and application. The first key phrase is "help others". The implication is that the duty of beneficence is one of cooperation more than coercion or control. The second key phrase is "their . . . interests". The implication is that the object of beneficence is other-regarding, not self-regarding. The interests and desires of the beneficiary take precedence, *ceteris paribus*, over the interests and desires of the benefactor. The third key phrase is "important and legitimate interests." This element directs attention to the evaluative dimension of the principle. The final phrase is also significant as it relates to the limits of the duty, but this aspect of the principle is beyond the consideration of this review. These emphases contribute to an understanding of the meaning and methods of beneficence. They also suggest points at which the principle is misunderstood and misapplied.

The end to which beneficence properly contributes has been described in many ways. The diversity of opinion regarding the proper object of beneficence is indicative of the disagreements that are possible when the constitutive elements of the principle are explicated. If the specific end or object of beneficence is not agreed upon, the methods of beneficence may be subject to an even greater, and perhaps more intense, disagreement. Suggestions of the proper end or object of beneficence range from happiness to self-determination.

John Stuart Mill viewed beneficence as part of a general policy that promotes individual and social utility. He held that the true object of beneficence is happiness which he understood as pleasure and the absence of pain ([19], pp. 14–15; [20], pp. 17–18). A contemporary rule-utilitarian, Richard Brandt, agreed with Mill that utility is the ultimate principle. However, he interprets utility in terms of welfare. Welfare, according to Brandt, is understood " . . . in the sense of liked experiences or activities, but not in the sense of the occurrence of events desired, as such, independent of their influence on liked occurrent states" ([4], p. 148).

Interestingly, Immanuel Kant used the same term as Mill, 'happiness', to designate the object of beneficence but interpreted it in a slightly different way. For Kant, happiness referred to well-being which, in his rendering, seems to mean the basics for bodily life. The principle and the duties related to it requires "making another's happiness one's end" ([15], p. 120) according to one's means to do so for another in need. The language of well-being as the proper end of beneficence also has contemporary reference. Gene Outka's discussion of *agape* as neighbor-love can be considered a theological parallel to philosophical beneficence. According to Outka, neighbor-love requires an "active concern for what *he* [the neighbor] may want or need, and not for the sake of benefits to the self [the agent]" ([22], p. 9). The object of this active concern is the neighbor's well-being ([22], p. 260). The basic contents of well-being can be identified and should be seen as inter-related; (a) putative value of the neighbor, (b) welfare – a blanket heading for interests important to physical survival, acquirement of skill and knowledge, affection and self-respect, *et al.*, (c) freedom – including conscious choices, deliberate decisions, and preferences. Freedom, according to Outka, is linked with reverence for the moral capacities of another ([22], pp. 263–274, esp. pp. 263–267).

The emphasis of John Rawls shifts away from external goods as the end of beneficence to an 'enriched quality of everyday life'. Rawls's definition of the duty of mutual aid is basically equivalent to the principle of beneficence. He stated the duty of mutual aid as "the duty of helping another when he is in need or jeopardy, provided that one can do so without excessive risk or loss to oneself" ([24], p. 114). This positive duty is followed by two negative duties: "The duty not to harm or injure another; and the duty not to cause unnecessary suffering" ([24], p. 114). The good or "the primary value of the principle [mutual aid] is not measured by the help we actually receive but rather by the sense of confidence and trust in other men's good intentions and the knowledge that they are there if we need them. Indeed, it is only necessary to imagine what a society would be like if it were publicly known that this duty was rejected" ([24], p. 339). Thus, the value of beneficence, in Rawls's theory, seems to go beyond mere reciprocity to something more intangible but vitally important to one's quality of life.

Alan Donagan suggested another variant of human well-being as the proper end of beneficence. He allowed for the principle to ordain a general policy for human activity without it cataloging any specific action or actions. The principle provides a policy for human flourishing or the "development and exercise of human potentialities" ([5], p. 85) by morally permissible means ([5], pp. 85–86). G. J. Warnock, similarly to Donagan, saw beneficence as one part of morality. The aim of morality, of which beneficence is a part, is to ameliorate the maleficent effects of 'limited sympathies'. In Warnock's system, the human predicament is such that things are inherently liable to go badly because of 'limited sympathies'. Thus, the end of beneficence and morality is to countervail the effect of limited human sympathies ([27], pp. 25–26, 85–86).

The final end or object to be mentioned here is self-determination. Willard Gaylin interpreted the proper end of beneficence on the basis of certain biological observations of humanity. He recognized the problems associated with basing moral theory on biology but suggested that acts of beneficence should be designed to relieve intrinsic and extrinsic dependencies, those internal and external deficiencies or impediments which inhibit one's self-reliance. He noted that, "whenever a person's sense of control over his own life is expanded; whenever he sees himself as the source of his own pleasures and security, his pride increases, his self-esteem increases, and his capacity for caring and concern is enhanced with it. The opportunity to sense one's self as a competent, independent, coping person allows one the generosity, unavailable in the humiliation of dependence, of sharing one's self with others and of exposing one's self to the vulnerability that loving inevitably implies. When an individual can be more self-reliant, fewer of his relationships have to be centered around the rather unreciprocal fulfillment of his own personal needs. He can proceed to service, to participation, to self-sacrifice, to love, to creativity, and to the caring for the intrinsically helpless" ([13], pp. 30–31).

These suggestions regarding the proper end of beneficence may point more or less toward the same state. Yet, the variety of conceptualizations and their basically formal character indicates something of the difficulty in interpreting the principle of beneficence. Each option may be a form of the good or benefit that beneficence intends without representing the complete, full, final, or only good or benefit.

The principle of beneficence can be expressed in a relativity simple and intuitively attractive statement, i.e., do good and prevent harm. It appears to set, as suggested by Donagan, a general policy for a moral community without defining specifically the proper end of that policy. Thus, differences of opinion result. The difficulties of stipulating the desired end of beneficence are consistent with and reflect the difficulties of stipulating the component elements of the principle.

III. ASSESSMENT OF GOOD

The principle of beneficence, by definition, enjoins the preservation or promotion of some good. Thus, the good to be realized in a given situation must be identified. This step is vitally important to a determination of the relevance and application of the principle in a given situation. Even though an assessment of a desired good is necessary, it is not always an easy task, nor is the assessment always agreed upon by the respective interested agents.

The notion that the moral community has a unified vision of the good or the good life is no longer held in modern Western societies. It is increasingly futile to appeal to such a notion as a shared context for moral reasoning, as justification for a given act, or as justification for establishing one view of the good life over another competing view. This situation may or may not be morally lamentable, depending upon one's tendencies to moral imperialism. Modern Western societies are now so pluralistic in nature that perhaps all hope of asserting or establishing a single, compelling vision of the good or good life which has normative status for the moral community seems lost or, at least, unlikely.

Accompanying a recession of the validity of a unifying, broad conception of the good, there has been an increased emphasis on the importance and adequacy of individual or personal conceptions of the good as a starting point of moral dialogue. These more individualistic notions are generally viewed as more limited in scope and authority, but important nonetheless to the moral life and a necessary consideration in moral judgment. Western democratic societies have de-emphasized the preservation and promotion of the common good as an over-riding value. Instead, the preservation and promotion of the good of the individual, which may beneficially impact on corporate life, enjoys superior status. Hans Jonas saw this when he noted that the idea of the primacy of the individual has achieved axiomatic status in the West ([14], p. 307).

This inverted order of priorities can rest on any number of premises. For example, one could argue that the good of and for an individual is more concrete, less abstract, than the good of and for society. Further, it could be held that the good of and for an individual is more subject to being known and less subject to diverse renderings than the good of and for society ([14], p. 305). One could also argue, along the lines of John Stuart Mill, that a preference for plural individual goods results in a greater gain for all than an establishment of a single 'common good'. Mill wrote, "mankind are greater gainers by suffering each other to live as seems good to themselves than by compelling each to live as seems good to the rest" ([19], p. 17). His appreciation for individual freedom did not rest only on his estimate of a beneficial result to all. More accurately, his was an appreciation of liberty itself and a qualified affirmation of the individual. He wrote, "If a person possesses any tolerable amount of common sense and experience, his own mode of laying out his existence is the best, not because it is the best in itself, but because it is his own mode" ([19], p. 82). The influence of Mill's liberalism on the modern endorsement of moral pluralism is rarely questioned.

An environment that tolerates and accepts plural perceptions of the good or the good life can be quite dynamic. The potential exists for elements of any two or more perceptions to agree. The opposite potential also exists; elements of any two or more perceptions may compete or conflict. The situation can become even more volatile once it is recognized that individual perceptions tend not to be immutable. One's perception is subject to change over time and is a function of numerous physical, historic, and social variables ([26], pp. 281–282). The potential for disagreement and debate about the identification and establishment of the good or good life seems limitless. Each perception may contain elements that are and some that are not supportable from a moral point of view. One might speculate that such a permissive atmosphere would result ultimately in a moral anarchy where each member functioned as a sovereign moral authority and concern for a common morality would perish. Perhaps fortunately, this has not been the case.

P. F. Strawson argued that some minimum common morality is necessary if plural perceptions of the good or good life are accepted or tolerated in a complex social organization. The core of this minimum common morality must include some vision of the good which admits of diversity in individual perceptions. Without a boundary or limit-setting minimum common morality, human society could not exist as it is now known. Strawson wrote, "some complexity of social organization seems, rather, a practically necessary condition of the ideal's [conflicting or competing individual ideal images or visions of human good or goods] being realized in any very full or satisfactory way. Now it is a condition of the existence of any form of social organization, of any human community, that certain expectations of behavior on the part of its members should be pretty regularly fulfilled: that some duties, one might say, should be performed, some obligations acknowledged, some rules observed. We might begin by locating the sphere of morality here. It is the sphere of the observance of rules, such that the existence of some such set of rules is a condition of the existence of a society. This is a minimal interpretation of morality. It represents it as what might literally be called a kind of a public convenience: of the first importance as a condition of everything that matters, but only as a condition of everything that matters, not as something that matters in itself" ([26], p. 284). Such a minimum common morality would allow individuals to select or construct and pursue their own perception of the good or good life so long as such activity did not traverse the boundaries which give definition or identity to the social organization. Not only would individuals be free to pursue their own ends, they would be free also to associate with others to constitute a sub-group of the social organization pursuing some shared or mutually agreed-upon ends. Again, the sub-group would be subject to the limitations or boundaries represented in a wider common morality. Thus, the importance of a wider common morality increases rather than decreases as individuality and plural perceptions of the good or good life are tolerated and safeguarded ([26], p. 296).

Any society, like modern Western democracies, that permits diverse perceptions of the good will likewise permit diverse normative judgments regarding the good which is by definition one end of beneficence. The analysis of interpretations of the good or goals of life by J. L. Mackie is instructive. He suggested that, "Moral reasoning consists partly in achieving a more adequate understanding of this basic good (or set of goals) [of human life], partly in working out the best way of pursuing and realizing it" ([18], p. 46). Moral reasoning about the goal and how it can be realized can take either of two radically different approaches, descriptive or prescriptive.

The descriptive approach involves an assertion of some end or state as the good or general goal of human life, which is to say simply that this is what humanity, or some representatives of humanity, in fact pursue or will find satisfying, or that if this end or state is postulated as an implied good, actual human activity becomes more intelligible and subject to a discovery of a coherent pattern in what would otherwise be a confusing constellation of chaotic and idiosyncratic purposes ([18], pp. 46–47). Mackie warned that the described good which results from this approach should not be understood to exclude a diversity of fundamental human purposes equally subject to some description. Neither should the sources of human satisfaction be interpreted as singular: there may be more than one source.

The prescriptive approach involves an assertion of some end or state as the good or general goal of human life which is to identify this end or state as proper to human life, that humanity or individuals ought to strive after it, whether or not it is presently being sought. Again, Mackie warned that this approach involves a subjectivity which reflects the speaker's view of the world that is probably based upon and consistent with certain fundamental goods or goals in the descriptive sense ([18], p. 47).

Given more than one way to reason morally about the good and its realization, the points of reference for claims of the good can be diverse and uncertain. What is common to these multiform constructions is the belief that what is termed good is represented to be 'such as to satisfy' some set of requirements, wants, or interests inherent to the particular construction ([18], pp. 55–63, esp. pp. 55–56). If these notions of the good or good life, and the preferred means by which to achieve or approximate it, are individually, contextually, and dynamically held, it seems reasonable to assume that no single end or state is sufficient to satisfy at once the requirements of such diversity. Thus, to hold an all-inclusive view of the good or goal of human life that is compelling and/or normative appears impossible in the modern West. If diversity is tolerated, even endorsed and protected, conflicts which turn on different notions of the good are inevitable.

The context of beneficent decisions and actions which aim at the good of the beneficiary can be quite complex if the preceding analysis is, in fact, accurate. Given an absence of a shared compelling vision of the good, together with a recognition of a human tendency to universalize one's particular perception, knowledge of the subject's (beneficiary) perception of his or her good appears to be a necessary condition for beneficence. If the subject's perception is ignored or over-ridden, the situation is perhaps one of callousness or paternalism but not simple beneficence. This is not to deny the potential problems in learning the subject's perception of the good, given his or her then present situation. In the best of worlds, moral agents would be able to articulate and defend, to the extent possible, their particular notion of the good. Yet, in a characteristically imperfect society, such as is reality, a person's view of the good may be expressed, and therefore learned by others, perhaps more indirectly "in their choice of words, their assessment of others, their conception of their own lives, what they think attractive or praise-worthy, what they think funny: in short, the configurations of their thought which show continually in their reactions and conversation" ([21], p. 202). Thus, instead of being able to rely on clear, rational, defeasible individual statements of the good, agents of beneficence may frequently rely by necessity on interpretations that are more subject to inaccurate or incomplete understanding.

In addition to a knowledge of the good as perceived by the subject, an agent must make several other assessments in order for beneficence to be a relevant principle in a given situation where some good is sought or at risk. Obviously, the situation must be perceived by the beneficent agent as one of want, risk, or deficiency. The want, risk, or deficiency may be seen as related to the subject's life or quality of life in terms of one's physical or mental condition. Food, shelter, clothing, protective custodial care, or medical treatment are examples of possible beneficent responses to a perception of need or want of this type. The want, risk, or deficiency also may be seen as related to the subject's social well-being or social opportunity in the community. Economic assistance to the poor, job training for the unemployed, payments to the jobless, and other types of social welfare are examples of possible beneficent responses to the latter type of perception. It seems that people are more apt to perceive the former rather than the latter as relevant to beneficence. Further, given a more evident perception, people tend to be more willing to respond in these situations with offers of some form of assistance. One can speculate that this greater propensity to identify with and respond to persons in the first instance may be related to a greater tendency and willingness to extend one's sympathy to those who might be considered blameless for and in their distressed situation. Other speculations are certainly possible.

Another assessment an agent must make is evaluative. The perceived good to be realized or the deficiency ameliorated must be seen as an end worthy of realization. This evaluation will be based, at least in part, on a background of values, tastes, and preferences of the agent. The agent's evaluative framework enables him or her to rank order the perceived relative goods, wants, risks, or deficiencies that are seen at issue. A judgment then can be made as to whether some beneficent response is indicated.

A third assessment for the agent is attainability. The agent will consider a range of possible courses of action and the range of likely consequences of these possible actions. Both the possible actions and anticipated consequences are assessed for their likelihood to achieve the desired end. The evaluative element also considers whether or not the desired end is commensurate with the cost, broadly construed, of the effort.

These additonal assessments, beyond a definition and identification of the good desired in a given situation, are merely suggestive of the sorts of considerations inherent to an end of 'the good' in beneficence. Given the variables related to 'the good' and certain limiting factors such as availability of resources, information, sympathy, and intelligence, plus the relative nature of the assessments necessary to beneficent conduct, it is reasonable to conclude that not all important and legitimate goods will be realized and not all risks, harms, or deficiencies will be avoided or lessened. A certain degree of misfortune, wastage, tragedy, or even wrong may well be an inevitable component of the human condition (cf. [27], pp. 17–22).

A few words of summary before considering a second definitional end of beneficence. The human condition is such as to generate, even require, multiple notions of the good or the good life. Not all of the perceptions are necessarily of equal value from a moral point of view. Yet, neither is one so compelling as to mandate its enforcement upon all people. There may be defeasible elements to several that are not mutually exclusive, though perhaps some aspects may be contradictory. A moral agent is called upon to make some judgment as to which among limitless options is a better vision or instance of the good or good life which is one object of beneficence. A moral agent is called upon to judge not only what the desired end or considered good is and how it can be achieved, he or she is required also to determine how it can be achieved or approximated in a morally licit fashion.

The possibilities of meanings and methods of the good appear almost endless. Definitive conclusions about these matters are profoundly elusive. The situation would be improved but not settled if a stipulative meaning of the good was agreed upon. The substantive or normative moral questions associated with it would still need to be addressed ([18], p. 63).

IV. ASSESSMENT OF HARM

In addition to good, the principle of beneficence, by definition, enjoins the prevention or decrease of harm to persons. This second imperative generally is considered more stringent than the imperative to do good. Some of the difficulties and uncertainties germane to an identification and realization of a good are parallel to some of the problems of knowing what constitutes harm and of selecting the moral means by which it can be avoided or lessened.

An identification of harm appears more subject to a general consensus of opinion than an identification of a good. The constitution of harm seems more universal and less individualistic, more concrete and less abstract, more objective and less subjective than the constitution of good. To some extent, the concept of harm appears more subject to definition since it might be construed as whatever the good is not. Harm could be that which lessens the good or departs from the norm for that entity.

An appeal might be made to human experience and intuition to support reasoning along these lines. Those things or conditions that produce pain or limit human satisfactions seem subject to being designated as harm or harmful. Those experiences that one interprets as harmful tend to be universalized as harmful to all people. When others are perceived to be similarly situated the potential exists for a sympathetic response, which may or may not result in a beneficent act, to be elicited. However, reservations and cautions similar to those voiced in regard to an identification of the good, goods, and goals of human life are appropriate to an identification of and response to harm. An application of one's experience to others may not be accurate and one's response or effort may not be appropriate, effective, or right.

A duty to lessen or avoid harm, represented in the principle of beneficence, leads to a consideration of its causes or enabling conditions. People seek an explanation and justification when harm threatens or eventuates. This is true also when it is believed that the harm could have been avoided and was not. People seek to control the negative result by an identification of its cause and by constructing means by or conditions in which it can be prevented or decreased. These interests to prevent or decrease are the inverse of those to promote the good discussed in the previous section. As there were many variables associated with the concept of the good in beneficence, so there are many variables associated with the notion of harm in beneficence and the moral options which exist to prevent or lessen it.

This discussion of harm could lead to an excursus on paternalism as a form or function of beneficence. For the purposes of this essay, paternalism is understood as a variant of beneficence which requires its own justification apart from and in addition to any appeal to beneficence. However, a consideration of preventing or lessening harm as a function of beneficence is appropriate here since by definition it is within the scope of the principle and these duties generally are seen as lexically prior to doing good. Thus, it will be profitable to discuss some of the difficulties of defining harm, its cause, and control in order to illustrate some of the problems in fulfilling one's duties of beneficence.

This analysis of harm sets aside the issues associated with the difference between conduct that affects only the agent and conduct that affects others. It seems that almost all conduct has some greater or lesser, direct or indirect impact upon others. Who is harmed and the potential limitations of one's duty to intervene in harmful or potentially harmful situations are issues that are not central to the conceptual inquiry undertaken here. This is not to suggest that they are unimportant or irrelevant to a thorough discussion of the principle of beneficence or paternalism. They are set aside in order to keep the present inquiry within manageable limits, i.e., to focus on the notion of harm as an element of the principle of beneficence.

An assessment of harm or potential for harm will reflect factual and evaluative factors of the situation in question. Joel Feinberg's discussion of harm is instructive ([9], pp. 106–113). He made several distinctions regarding harm as a basis for action. The first distinction is between 'harm and wrong'. Some things may harm but not be wrong. For example, drinking a lethal dose of arsenic will kill a thirst and the one who drinks it, but if it is swallowed knowingly, voluntarily, and deliberately, the individual would not be necessarily wronged by the action. Also, some things may harm and be wrong. For example, swallowing a lethal dose of arsenic will harm, and if it is poured down the throat of an objecting, restrained person, the harmed individual would also be wronged.

A second distinction is between 'harm and risk of harm'. Some actions are certain to produce harm and the harm is the desired end of the action. Other actions create a risk of harm and the potential harm is not necessarily desired. It may be a by-product of some conduct directed toward another end. For example, a person cuts off a hand because it offends him or her, producing a direct and deliberate harm. Another person drives a car at high speeds in order to reach his or her destination sooner and produces an increased risk of harm to the agent and others.

A third distinction is between 'reasonable and unreasonable risks of harm'. All human activity engenders some risk of harm or injury. Judgments are made on the basis of selected criteria as to whether or not these risks are reasonable (cf. [9], pp. 109–110). No simple calculation of risk and benefit can decide reasonableness. For example, such an assessment would include the probability and magnitude of harm, probability and importance of the desired end, and the necessity of the means. "But in a given difficult case, even where questions of 'probability' are meaningful and beyond dispute, and where all the relevant facts are known, the risk decision may defy objective assessment because of its component personal value judgments" ([9], p. 110).

A fourth distinction is between 'fully voluntary and not fully voluntary assumptions of risk'. Here it should be noted that voluntariness is a matter of degree. Those actions or assumptions of risk that are more nearly fully voluntary have their origin in the agent and express the agent's decided values and preferences. Those that are less than fully voluntary do not have these characteristics.

Which factors in a situation occasion beneficence: harm or wrong, actual harm or risk of harm, the degree of reasonableness or voluntariness of risks of harm? Is beneficence called for by the fact of harm or merely the conditions of harm about which certain evaluative opinions are held?

If one desires to lessen or prevent harm, consideration must be given to an identification of the causal factor or factors in a harmful situation by which it can be controlled or ameliorated. These harmful situations can be complex. A judgment regarding the means by which the deleterious element can be controlled involves a degree of selectivity. Again, a discussion by Joel Feinberg of three relative factors in assigning causality and in identifying a means by which the resultant harm can be controlled is helpful ([8], pp. 143–148).

The first form of causal relativity is relativity to what is 'usual or normal' in a given situation. Departures from or additions to that which is considered normal or usual may be chosen as a necessary condition for a harm to ensue. A selection of 'the cause' can be based on an understanding of what is 'normal', and 'normal' varies from situation to situation.

A second form of causal relativity is 'relativity to ignorance'. When causes are sought one is attempting to render the considered event intelligible. "Intelligibility, however, is always intelligibility *to* someone, and understanding is always *someone's* understanding, and these are in part functions of what is already known or assumed to be normal or routine" ([8], p. 144).

The third form of causal relativity discussed by Feinberg is 'relativity of practical interest'. Agents tend to identify the cause of an event with the condition which seems subject to a manipulation or alteration that will change the harmful course of events.

This brief discussion and review of some relative factors in causation suggests that the meaning of 'cause' can vary. Feinberg's cogent summary deserves quoting. "Those who are concerned to produce something beneficial seek 'the cause' of what they wish to produce in some new condition which, when conjoined with the conditions usually present, will be *sufficient* for the desired thing to come into existence. On the other hand, those whose primary aim is to eliminate something harmful are for the most part looking for causes in the sense of *necessary* condition. That is because, in order to succeed in such a task, one must find some condition in whose absence the undesirable phenomenon would not occur and then must somehow eliminate *that* condition. Not just any necessary condition, however, will do as 'the cause'; it must be a necessary condition which technicians can get at, manipulate, modify, or destroy. Our purpose here determines what we will accept as 'the cause,' and . . . accessibility and manipulability are as important to our purposes as the 'necessity' of the condition" ([8], p. 145). Thus, the relative interests and purposes of agents in a given context are an implicit part in the selection of 'the cause' of an event. The 'discretionary character' and 'contextual relativity' of identifying 'the cause' of harm suggests that a beneficent act may be difficult to construct, identify, execute, and be consistently described as such.

These discretionary, evaluative, and relative features of an identification and amelioration of harm indicate how and why harm is subject to diverse interpretations. An effect of this sort of interpretive uncertainty is a potential for disagreement about the relevance and application of the principle of beneficence in a given situation. The issue becomes twofold: (1) what constitutes harm?, and (2) when does harm become sufficient for beneficence? Opinions can vary about a single situation and from one context to another. The opposing positions regarding euthanasia taken by Marvin Kohl and Arthur Dyck illustrate how harm and its meaning for beneficence can be subject to different understandings.

Marvin Kohl [16] understood harm in terms of the pain and suffering of a terminally ill patient. He considered the principle of beneficence to permit killing terminal patients who meet certain necessary and perhaps sufficient conditions in order to lessen the harm to that person of his or her pain and suffering. In other words, Kohl considered the harm of pain and suffering as greater than the harm of killing persons. The act of killing would be described as beneficent euthanasia.

Arthur Dyck [6] opposed Kohl's stance on killing terminal patients as an instance of beneficence. Dyck held that the harm of killing is greater than the harm of pain and suffering. He emphasized the prescription in beneficence that one ought to prevent or lessen harm, not compound the harm by an additional harmful act of killing. He did not accept Kohl's reasoning that beneficence allows harm in order to lessen harm in the case of terminally ill patients. Thus, the two commentators disagreed on reasonable moral grounds about the meaning of harm, its relative significance to beneficence, and the means by which it can be controlled.

The variables, some of which have been discussed, within an interpretation, identification, and control of harm necessarily allows for a diversity of opinion regarding the meaning and application of the principle of beneficence to complex situations that are perceived to produce or permit harm. An intuition that harm is more subject to a consensus identification and method of relief is inaccurate. Understandings of harm can be as diverse as the possible perceptions of the good or the good life. The variable character of harm, like that of the good, does not mean that one interpretation is accurate and all others are in error. Each may simply reflect an individual's ordering of the relative factors inherent to an understanding of a complex situation. Thus, an application of the principle of beneficence is relative to a discretionary and evaluative assessment of harm and its cause.

V. QUALITIES OF BENEFICENCE

In addition to the ends of good and lessened harm, the principle of beneficence recommends conduct with specific qualities or characteristics. Three specific qualities are presented here as central to an understanding of the duties of beneficence interpreted along the lines of Beauchamp and Childress. This is not to suggest that actions without these qualities cannot effect good or decrease harm. It is to suggest, however, that these qualities endow certain beneficial acts with a distinctive character appropriate to the principle of beneficence as distinguished from other principles like paternalism.

One central quality of beneficence is a respect for persons. The principle of beneficence affirms the status of persons as valued and worthy of assistance in attaining *their* legitimate ends or relieving *their* distress or dependency. Conduct appropriate to beneficence characteristically does not disregard the rights and freedoms of the beneficiary. Immanuel Kant called attention to this dimension of beneficence. Beneficence does not allow for the self-exaltation of the benefactor or a disregard of the recipient's dignity. The end of the subject is the proper intent of beneficence, not the benefactor's end of which the beneficiary may be a means ([15], p. 116).

This sort of restraint or limit is a necessary condition of pure beneficence. The primary subject of beneficence is the weal and woe of the recipient, not the provider. To accept this sort of limiting condition as necessary to beneficence means that a beneficent agent may permit, out of respect, another to err. It does not mean that the agent must assist another to attain ends or goals that are not held to be worthy ([4], p. 147).

John Rawls's discussion of mutual respect is applicable to what is referred to here as respect for persons. For Rawls, mutual respect is a duty that is owed to individuals. Respect is shown to others in several ways: "in our willingness to see the situation of others from *their* point of view, from the perspective of *their conception of the good*; and in our being prepared to give reasons for our actions whenever the interests of others are materially affected" ([24], p. 337, emphasis added). Respect also includes "a willingness to do small favors and courtesies, not because of any material value, but because they are an appropriate expression of our *awareness of another person's feelings and aspirations*" ([24], p. 338, emphasis added). The quality of respect for persons expressed in these and other ways is appropriate to acts of beneficence which intend the legitimate ends of another person.

A second central quality of beneficence is intimately related to respect for persons. A regard for the self-esteem of the beneficiary is a logical corollary to a respect for persons. A regard for self-esteem signals that self-esteem is an important good which should be considered in fashioning one's benefaction. Again Rawls's remarks are insightful. He identified self-esteem as "perhaps the most important primary social good" in his system of thought ([24], p. 440). Self-esteem, according to Rawls, has two aspects. One includes "a person's sense of *his own* value, his secure conviction that *his* conception of *his* good, *his* plan of life, is *worth* carrying out" ([24], *ibid.*, emphasis added). A second aspect "implies a confidence in *one's ability*, so far as it is within one's power, to fulfill *one's intentions*" ([24], *ibid.*, emphasis added). An orientation to and appreciation for another's considered judgments, values, and goals, represented as a regard for self-esteem are applicable to the principle of benefience.

Since beneficence is based on the putative value or worth of an individual, a regard for self-esteem becomes a necessary condition. It requires, at least, that one's benefaction not emphasize unnecessarily the recipient's deficiency or dependency; that it take a form of assistance rather than an obstruction to legitimate self-determination; and that it take a form of cooperation with rather than coercion or control of the beneficiary. Other specific instructions on the methods by which self-esteem is regarded are possible. These are suggested in order to illustrate some of the qualitative considerations in constructing one's benefaction.

A third, and final, quality to be noted is the value of freedom and liberty.

Again, this quality is inter-related to the prior two, each consistent with and inter-dependent upon the other. H. Tristram Engelhardt included valuing freedom among those goods that many would envisage as central to a full moral life ([7], p. 12).[1] If this assertion is accurate and justifiable, which seems true, then freedom will be necessarily valued and respected in a moral community that acknowledges plural perceptions of the good and the good life.

One of the greatest champions of individual liberty, John Stuart Mill, noted that an endorsement of liberty does not require an unconcern for the good, well-being or well-doing, of others. Valuing liberty does mean, according to Mill, that "disinterested benevolence can find other instruments to persuade people to their good than whips and scourges, either of the literal or metaphorical sort" ([19], p. 92). Mill held that individuals who are mature in their faculties are sovereign over their own body and mind. He recognized that mature individuals might make unfortunate or even fatal errors in judgment and conduct. Yet, the freedom or liberty of a mature person to formulate and execute his or her life-plan should be respected, "not because it is the best in itself, but because it is his own mode" ([19], p. 82, cf. p. 13).

It should be remembered that this essay assumes situations involving mature persons and conduct that affects only the actor in any material way. The debates about the limits of liberty or autonomy are set aside. The concern here is to signal the place of liberty or freedom in beneficence. A concern to respect and value individual freedom and liberty is befitting to an interpretation of beneficence which considers as primary the beneficiary's acceptable vision of the good, the good life, and the means by which to achieve or approximate it. Beneficence would then necessarily include a consideration of the liberty or freedom of the recipient and seek to preserve and promote it.

How these qualities (respect for persons, regard for self-esteem, and value of liberty and freedom, among possible others) are understood, ranked, and expressed will influence the character of beneficent conduct. The form and character of beneficence may determine if it will be received by the proposed beneficiary and thus afforded an opportunity to fulfill its intended purpose.

These qualities are necessary and limiting conditions of beneficence. They, in part, enable beneficence to be distinguished from alternate forms of other-regarding conduct, e.g., paternalism. The moral value of beneficent acts is stronger when these qualities are honored and considered central to the beneficent act. The moral value is weaker when they are dishonored and considered optional or peripheral to a beneficent act. In either instance, the act may be effective or not. The relative moral value of the beneficent act would turn in part on the priority given to these necessary conditions or qualities.

VI. BENEFICENCE IN HEALTH CARE

The foregoing analysis shows that the general statement of the principle of beneficence by Beauchamp and Childress is superior to Frankena's because of the former's clearly expressed orientation to the defeasible good or goods as perceived by the intended beneficiary. This orientation is justified in part for several reasons. It accommodates diverse meanings of the good or the good life which do not submit to a single compelling vision. It accommodates the relative and contextual nature of an identification and assessment of harm, as well as a recognition of the perceptual and discretionary differences regarding its control. Finally, it accommodates the necessary and limiting qualities or conditions of beneficence which suggests that it is characteristically a cooperative and permissive, rather than a coercive and autocratic, enterprise.

An orientation of beneficence "to *help* others further *their* important and legitimate interests" ([2], p. 136, emphasis added) appears well-informed. The promotion of good and the prevention or decrease of harm provide objectives for beneficence but do not stipulate the content of these ends or the methods by which they are realized. In addition, the selected elements or qualities of beneficence discussed above are subject to seemingly limitless interpretation, application, and expression in complex situations. A recognition of the dynamic nature of these elements in beneficence means that disagreements can reasonably attend judgments of the relevance of the principle to particular situations and the action indicated by it.

An application of the principle of beneficence to health care can be particularly troublesome and problematic. Recall the statement of the American Medical Association that began this essay. The principal objective of the medical profession, and by extension to health care in general, "is to render service to humanity with full respect of the dignity of man" ([1], p. 39). This statement seems to reflect the thrust of the principle of beneficence, as it has been explicated above, with the language of service or benefit and respect. Both ends and associated duties are presented as role-related. They appear accentuated for the medical profession perhaps because of their special training, skills, and commitments. This emphasis is consistent with the more general duty of all moral agents to render or offer respectful assistance to persons in some form of distress or seeking some legitimate good. However, a danger exists in too great an emphasis on role-related duties. An over-emphasis on the duty to serve or benefit may obscure the corollary and equal component duty to respect.

As the moral community has changed to accept plural notions of the good, so has the environment of health care changed in which beneficence was an unquestioned justifying principle for decisions and actions. No longer is the therapeutic relationship viewed as one in which the interests of the parties are identical, compatible, or even complementary. They now may be viewed as contradictory and competing. It is now possible, but not necessary, to fashion the therapeutic relationship as an adversarial encounter in which trust has been usurped by suspicion. The discretionary authority formerly held by the care provider is no longer absolute. Encroachments on this preserve have taken form as patients' rights that limit certain professional liberties. A presumption that health care is offered by benevolent persons has fallen victim to a belief of their malevolence or disregard against which a defense must be waged. This radically changed situation in health care may be difficult for physicians and other providers to accept or accommodate. It is possible that when offers or efforts to benefit, in the name of common understandings of beneficence, are rebuffed by patients, physicians and others may think that they are rendered impotent to do that which they are charged to do, i.e., serve. They may sense that they are being denied the opportunity to do that which has been and is considered central to their professional identity and duty.

Medical education has taught physicians to understand good in terms of health and harm as disease or disability. We are learning that health and disease are as subject to varying definitions as are good and harm. Thus, the problems inherent to beneficence are compounded when applied to contexts of health care. If a broader scope for good and harm is reasonable, i.e., beyond the body to include social and other considerations, the competency of a physician to make these assessments and require a decided course of action is questionable.

Further conflicts may result because of medical training. Physicians are taught a limited number of treatment modalities to achieve a medical good. Yet, the moral environment of the broader community endorses pluralism and tolerance regarding paths to health, even if these paths are considered ineffective or harmful (e.g., laetrile for adults). To acknowledge this permissive atmosphere is not to say that all values, ends, and means are equally good or proper.

Patients increasingly are active participants in the therapeutic encounter.

Sally Gadow noted that a medical definition of medical benefit may conflict with a patient's autonomy in which case both cannot be satisfied at once ([12], pp. 683–684). The impression is that the duty to serve yields totally to the duty to respect when the provider's definition of service is not accepted by the recipient. But, the duty to serve, grounded in the principle of beneficence, does not necessarily cease in these situations of conflict. More accurately, it takes a different form when the patient's judgments do not require the physician to violate his or her own moral commitments. Indeed, right conduct may not conform to standard medical understandings of benefit. Edmund Pellegrino and David Thomasma have accurately stated that, "the action chosen must be the *right* one for this patient. This is to say, it must be as congruent as possible with his or her particular clinical context, values, and sense of what is 'worthwhile' or 'good.' What the physician and the patient seek together is a judicious decision, one which optimizes as many benefits and minimizes as many risks as the situation will allow" ([23], p. 124). Right decisions therefore will vary among different clinical situations.

Medicine is not totally unable to accommodate diversity. Engelhardt observed that medicine offers people a chance to gain control over their bodies and their futures ([7], p. 24). Each medical intervention presupposes some vision of the good and the goods of human life. Medicine can cooperate with and further ends which are contradictory but perceived as valid, e.g., birth control and fertility treatments. Thus, there may be unlimited possibilities for beneficence in health care. Plural ends and means may defy harmonization, but not validation necessarily. Each may have potential for respect, accommodation, and assistance by medicine without necessarily violating the moral commitments and moral integrity of the agents.

Beneficence in health care prescribes a general policy or objective without specifying how exactly it is to be achieved. The notion that service should be more highly valued than improvement is well-taken. Service implies respect for the perceived good of the subject. Improvement risks an imposition of the provider's perception of the good on the weaker party. Progress or the good cannot be demanded, even by physicians; it must be received as a grace (cf. [14], pp. 308–309).

Beneficence in health care seems to require the virtue of wisdom. According to Philippa Foot, wisdom has two parts. "In the first place the wise man knows the means to certain good ends; and secondly, he knows how much particular ends are worth" ([10], p. 5). The perceptual and evaluative elements of beneficence would be judiciously informed by an exercise of wisdom. If the intended good or intended decreased harm does not result because the effort is refused, or resources are unavailable, or the effort is not effective, the appropriate response per beneficence includes sorrow with a hope that the situation will change. The concern for and commitment to the threatened or deprived person that is expressed in the attempt or offer is a good to the one in peril, the one who offers, and to the moral community (cf. [3], pp. 14–15, 156).

Forms of beneficence in health care must be flexible and adapted to unique circumstances. Medicine has not been reluctant to champion the unique character of each medical event. It does not appear impossible for or a violation of medical practice to accept the fact that beneficence may take diverse tangible and intangible forms. The agents may preserve their relationship on the basis of their agreements and work toward an accommodation or resolution of their disagreements. The question of David Rothman is critical: Can the respective powers or rights be controlled and egos restrained enough to preserve the goods which ensue even though they may not be considered the preferred or ideal ones according to the criteria of the respective agents. (cf. [25], p. 95)?

The qualities of beneficence presented above as limiting conditions imply a willingness to accept responsibility *with* others without depriving others of their right to be fully responsible. As there is not only one way to harm, so there is not only one way to benefit and respect. Moral principles provide guidance for the moral life. Morality offers principles to live by, it does not detail how one should live (cf. [27], pp. 87–92). Medicine or health care should not attempt to do more than morality. Its capacities are finite. Its principles of conduct are derived and adapted to its particular activity. They are a part of and reflect morality in general. The enterprise of health care functions within that context and is constrained by it where necessary. An interpretation of beneficence, like the one provided here, does not constrain health care needlessly or immorally. Nor is its relevance to health care neglected. The principle of beneficence presents a formidable challenge to health care to do that which it holds as its primary objective in a morally licit fashion. The two specific charges embodied within the principle, to benefit and to respect persons, are mutually reinforcing, not mutually exclusive or contradictory in situations involving mature persons.

Institute of Religion and
Baylor College of Medicine
Houston, Texas

NOTE

[1] Engelhardt differentiates between respect for freedom as a value and respect for freedom as a side constraint. He sees the latter as more important and more easily justified than the former. Respect for freedom as a side constraint functions, according to Engelhardt, as a condition of the moral life and part of the 'hard core' of ethics. Valuing freedom is part of the 'soft core' of ethics, along with sympathy, charity, etc. . . . Cf. [7].

BIBLIOGRAPHY

1. American Medical Association: 1957, 'Principles of Medical Ethics (1957)', in S. J. Reiser, *et. al.* (eds.), *Ethics in Medicine*, The MIT Press, Cambridge, Mass. (1977), pp. 38–39.
2. Beauchamp, T. L. and Childress, J. F.: 1979, *Principles of Biomedical Ethics*, Oxford University Press, New York.
3. Blum, L. A.: 1980, *Friendship, Altruism, and Morality*, Routlege and Kegan Paul, London.
4. Brandt, R. B.: 1979, *A Theory of the Good and the Right*, Oxford University Press, New York.
5. Donagan, A.: 1977, *The Theory of Morality*, University of Chicago Press, Chicago.
6. Dyck, A.: 1975, 'Beneficent Euthanasia and Benemortasia: Alternative Views of Mercy', in M. Kohl (ed.), *Beneficent Euthanasia*, Prometheus Books, Buffalo, New York, pp. 117–129.
7. Engelhardt, H. T., Jr.: (not dated), 'Autonomy and Biomedicine', Unpublished manuscript.
8. Feinberg, J.: 1970, 'Action and Responsibility', in *Doing and Deserving*, Princeton University Press, Princeton, New Jersey, pp. 119–151.
9. Feinberg, J.: 1971, 'Legal Paternalism', *Canadian Journal of Philosophy* 1 (September), 105–124.
10. Foot, P.: 1978, *Virtues and Vices and Other Essays in Moral Philosophy*, University of California Press, Berkeley.
11. Frankena, W. F.: 1973, *Ethics*, 2nd ed., Prentice-Hall, Inc., Englewood Cliffs, New Jersey.
12. Gadow, S.: 1980, 'Medicine, Ethics, and the Elderly', *The Gerontologist* **20**, 680–685.
13. Gaylin, W.: 1978, 'In the Beginning: Helpless and Dependent', in W. Gaylin, *et al.*, *Doing Good: The Limits of Benevolence*, Pantheon Books. New York, pp. 12–38.
14. Jonas, H.: 1969, 'Philosophical Reflections on Experimenting with Human Subjects', in S. J. Reiser, *et al.* (eds.), *Ethics in Medicine*, MIT Press, Cambridge, Mass. (1977), pp. 304–315.
15. Kant, I.: 1964, *The Doctrine of Virtue: Part II of the Metaphysic of Morals*, Mary J. Gregor, trans., University of Pennsylvania Press, Philadelphia.
16. Kohl, M.: 1975, 'Voluntary Beneficent Euthanasia', in M. Kohl (ed.), *Beneficent Euthanasia*, Prometheus Books, Buffalo, New York, pp. 130–141.
17. Maclntyre, A.: 1977, 'Utilitarianism and Cost-Benefit Analysis: An Essay on the Relevance of Moral Philosophy to Bureaucratic Theory', in K. Sayre (ed.), *Values*

in the Electric Power Industry, University of Notre Dame Press, Notre Dame, Indiana, pp. 217–237.
18. Mackie, J. L.: 1977, *Ethics: Inventing Right and Wrong*, Penguin Books, New York.
19. Mill, J. S.: 1956, *On Liberty*, C. V. Shields (ed.), Bobbs-Merrill Educational Publishing, Indianapolis.
20. Mill, J. S.: 1971, *Utilitarianism*, S. Gorovitz (ed.), Bobbs-Merrill Company, Inc., Indianapolis.
21. Murdoch, I.: 1966, 'Vision and Choice in Morality', in I. T. Ramsey (ed.), *Christian Ethics and Contemporary Philosophy*, SCM Press, London, pp. 195–218.
22. Outka, G.: 1972, *Agape: An Ethical Analysis*, Yale University Press, New Haven.
23. Pellegrino, E. D. and Thomasma, D. C.: 1981, *A Philosophical Basis of Medical Practice*, Oxford University Press, New York.
24. Rawls, J.: 1971, *A Theory of Justice*, Harvard University Press, Cambridge, Massachusetts.
25. Rothman, D. J.: 1978, 'The State as Parent: Social Policy in the Progressive Era,' in W. Gaylin, *et al.*, *Doing Good: The Limits of Benevolence*, Pantheon Books, New York, pp. 67–96.
26. Strawson, P. F.: 1966, 'Social Morality and Individual Ideal', in I. T. Ramsey (ed.), *Christian Ethics and Contemporary Philosophy*, SCM Press, London.
27. Warnock, G. J.: 1971, *The Object of Morality*, Methuen and Co., Ltd., London.

JAMES F. CHILDRESS

BENEFICENCE AND HEALTH POLICY: REDUCTION OF RISK-TAKING[1]

I. THE PRINCIPLE OF BENEFICENCE

In Alison Lurie's *The War Between the Tates* ([18], p. 271), a character says, "I was less morally ambitious than you; I didn't aspire to do good. I only wanted not to be harm." We commonly distinguish between doing good and not doing harm, the latter appearing to constitute the moral minimum in human interactions. Philosophers sometimes use 'beneficence' to refer to doing and 'nonmaleficence' to refer to not doing harm. Although not fully adequate, these terms are useful for shorthand purposes. I shall also distinguish 'beneficence' and 'benevolence', using 'benevolence' to refer to the will or disposition to benefit others, and 'beneficence' to refer to the act of benefitting others.

There are both broad and narrow definitions of beneficence. The broad definition encompasses nonmaleficence; doing good includes not doing harm. There is warrant for this broad definition, because anyone who violates the duty of nonmaleficence cannot be said to have discharged the duty of beneficence. But ambiguities remain, for the debate about 'beneficent euthanasia' is precisely whether death is always a harm whose infliction is maleficent, and if it is a harm whether it can be outweighed by benefits such as the elimination of pain and suffering. One reason for distinguishing nonmaleficence and beneficence is that doing good sometimes seems to conflict with not doing harm in relation to the same person (e.g., euthanasia) or to different persons (e.g., non-therapeutic research). And in cases of conflict, the duty of nonmaleficence, *ceteris paribus*, has priority.

Two recent broad definitions of the principle of beneficence are worth attention. The first is the Belmont Report of the National Commission for the Protection of Human Subjects of Biomedical and Behavioral Research [20] which distinguishes three basic ethical principles – respect for persons, beneficence, and justice. In ways that are not always clear or cogent, the National Commission maintains that the principle of respect for persons entails both respecting autonomous persons and protecting nonautonomous persons from harm, while beneficence involves "efforts to secure their well-being." According to the National Commission, two general rules are

Earl E. Shelp (ed.), Beneficence and Health Care, 223–238.

complementary expressions of the principle of beneficence in research involving human subjects: (1) do no harm and (2) maximize possible benefits and minimize possible harms. These rules also require risk-benefit analysis. In applying its short list of principles, the National Commission could and should have considered the protection of nonautonomous persons from harm as a requirement of beneficence. It is more plausible to view this protective rule as derivative from beneficence than as derivative from respect for persons. For if, under beneficence, we are to do no harm and to maximize benefits and minimize harms, it is clear, at least by implication, that we should protect people who cannot protect themselves from harm. The problems in the National Commission's formulation are not limited to the application of principles, to their expression or specification in rules, but extend to the principles themselves. It is more plausible to hold that 'do no harm' is a distinct moral principle and that both the principle of beneficence (produce benefits) and the principle of nonmaleficence (do no harm) generate the requirement to produce the greatest possible balance of good over evil, or benefit over harm, in an imperfect world. This requirement is the principle of utility, which is "a heuristic maxim in conflict situations" – to use William Frankena's phrase – involving benefits and harms ([12], p. 48). It is clearer and more adequate to distinguish five principles: beneficence, nonmaleficence, utility, respect for persons, and justice [4].

Likewise offering a broad definition of beneficence, William Frankena distinguishes four elements: (1) One ought not to inflict evil or harm (what is bad), (2) One ought to prevent evil or harm, (3) One ought to remove evil, and (4) One ought to do or promote good ([12], p. 47). Frankena has arranged these elements in order of priority: the first is more stringent than the second, and so on. And he admits that the fourth – one ought to do or promote good – may be an ideal rather than a duty. Even such a broad definition has to encompass distinctions that are important for ordinary moral reflection; for example, the Hippocratic principle *primum non nocere* (first of all, or at least, do no harm) presupposes the distinction between not harming and doing good and gives the former priority over the latter (which it also recognizes in the principle of 'benefitting the sick').

A more satisfactory approach is to recognize these distinctions at the outset in order to clarify the different weights of principles and their independence as well as their possible conflicts. Using Frankena's categories, I would propose the following classification:

Nonmaleficence: 1. One ought not to inflict evil or harm (what is bad)
Beneficence: 2. One ought to prevent evil or harm

3. One ought to remove evil or harm
4. One ought to do or promote good.

Imperative No. 1 involves noninfliction of harm. Imperatives Nos. 2–4 involve positive actions for individuals or groups: preventing and removing harm and promoting good.

The principle of beneficence applies to groups as well as to individuals, whether we focus on the agents of beneficence or on the objects of beneficence. For example, the society may have a duty to protect its weak and vulnerable members. Likewise, the targets of beneficent acts may be groups, even 'statistical persons,' as well as 'identified persons'. It is not my task in this chapter to defend the duty of beneficence, which may be independent or grounded on or at least supported by other principles such as justice ([4], Chapter 5); my task is rather to explore the *implications* of the principle of beneficence for health policy.

II. PREVENTION OF ILL HEALTH AND EARLY DEATH

We could examine several different areas of health policy in terms of the principle of beneficence. Consider, for example, various governmental responses, actual and proposed, to health needs: provision of medical care to the needy, coverage of the costs of medical care, prevention of ill health, and promotion of good health. The principle of beneficence, and perhaps other principles such as justice, may justify all these policies, but when resources are limited, hard choices may be required. Indeed, it may even be necessary to decide whether health should have priority over other basic needs and desires. Benjamin Rush, an eighteenth-century physician, contended that "there is a grade of benevolence in our profession much higher than that which arises from the cure of disease. It exists in exterminating their causes" ([13], p. 525). Whether, however, this expression of benevolence should have priority over rescue or crisis interventions will depend on several factors, including costs, effectiveness, and symbolic significance. Because I have examined these priority questions elsewhere [8], I will concentrate on the implications of the principle of beneficence for a governmental policy of prevention of ill health and early death.

Why concentrate on the prevention of ill health rather than the promotion of health? Within Frankena's categories of beneficence, promotion of good is a less stringent requirement than prevention of evil or harm. Numerous conceptual and normative questions are involved in analyses of health and disease. But however these are resolved (e.g., whether health should be defined as the absence of disease), the issues in governmental policy should be analyzed first of all in relation to prevention. If the arguments do not support a policy of prevention, they will not support a policy of promotion.

'Prevention', of course, is ambiguous. It is possible to distinguish primary, secondary, and tertiary prevention. Primary prevention consists of actions to avoid disease or ill health prior to its manifestation or occurrence in a particular individual. Secondary prevention involves detection of a disease before its symptoms have become evident and intervention to stop it or slow it down. An example is screening for asymptomatic hypertension and trying to control it; another example is screening for phenylketonuria in newborns and then instituting a dietary regimen. Tertiary prevention, which converges with conventional therapy, occurs after the disease or ill health has become clinically manifest and attempts to stop or slow its progression. An example is treating a person who has pernicious anemia with Vitamin B-12 in order to prevent its recurrence (for these distinctions, see [22], pp. 2, 21–22).

Secondary and tertiary prevention obviously involves health care professionals who can detect a disease prior to or after its clinical manifestation. Primary prevention, our major interest in this essay, includes strengthening individuals to make them less susceptible to disease and ill health (e.g., vaccinations), altering the physical or socioeconomic environment (e.g., reduction of pollution), and changing individual lifestyles and behavioral patterns (e.g., diet and exercise).

In the last two centuries prevention's successes have altered the causes of morbidity and mortality, and prevention now faces different obstacles.

> In the United States during the past century, a major shift in the nature of health problems has occurred – from a burden imposed primarily by infectious diseases (the leading killers in 1900 being pneumonia, influenza, tuberculosis, and combined diarrhea and enteritis) to a burden imposed by chronic diseases, accidents and violence. Many of the chronic diseases arise out of changes in the environment and personal life-styles – exposure to new environmental pollutants, increased stress, decreased physical activity, and increased consumption of certain foods, cigarettes, alcohol, and other substances. . . . Considering the population as a whole, today's major medical problems are the chronic diseases of middle and later life: heart disease, cancer, and stroke. For the population under 44, the leading causes of death, in order, are: accidents, heart disease, cancer, homicide, and suicide. For persons under 25, accidents are by far the most common cause of death, with homicide and suicide the next leading causes ([22], p. 19).

Many, although by no means all, of these causes of morbidity and premature mortality can be attacked through alterations in lifestyles and behavioral patterns, and I shall concentrate on this form of primary prevention without implying that it should have priority.

III. LIFESTYLES AND RISKS

In *The Republic*, ([28], p. 205), Plato's Socrates argues for a simple lifestyle – no sauces and sweets, no Attic pastry, and no Corinthian girls. His argument is, in part, aesthetic: "And to need doctoring, isn't that ugly – except for wounds, or the attacks of some seasonal illnesses?" To need medical care "because of sloth or the manner of life" is ugly.

According to several studies, an individual's lifestyle is a key determinant of his or her health [5, 6, 15, 22]. Consider the increase and the decline in coronary heart disease mortality. For the 1930's through the 1950's coronary mortality increased before it reached a plateau in the 1960's. From the late 1960's to the present it has declined by approximately 2.5% annually. A major cause of the decline appears to be dietary changes (especially less consumption per capita of animal fats and oils, butter, cream, and eggs) and decreased cigarette smoking especially among male adults [7, 21, 29]. Numerous other examples could be mentioned from such areas as physical activity, diet, psychosocial stress, accidents, violence, and the use of drugs, alcohol, and tobacco. Our lifestyles and behavioral patterns create the risk of morbidity and early mortality.

Life is inherently risk-filled, if we understand 'risk' as a chance of injury or loss. In ordinary discourse 'risk' refers to the *amount* of possible loss and to the *probability* of that loss. We both run risks and impose them on others. It is important to distinguish between *risk-taking* and *risk-imposition* [9]. The former refers to conduct that creates risks for the actor, while the latter refers to conduct that imposes risks on others. One important issue, to be considered later, is whether much or most individual risk-taking also imposes risks or, more generally, costs on others. 'Risk-taking' implies that the actor is aware of the risks and voluntarily assumes them. He may, of course, engage in 'risky' action without 'taking risks' in this sense.

A second and related distinction is between *voluntary* and *involuntary* risks. Often we voluntarily assume risks because we desire certain benefits, either for ourselves or for others (e.g., eating rich foods). But risks are also imposed on us against our wills (e.g., by reckless drivers). Apparently we are willing to tolerate more risk in voluntary activities, over which we have some control, than in involuntary activities [25].

Each person has what Charles Fried calls a *life plan* consisting of ends and values [14]. It also includes a *risk budget*, for we are willing to run certain risks to our health and survival in order to realize some ends and express some values. As Fried suggests, "a person's life plan establishes the magnitudes of risk which he will accept for his various ends at various times in his life" ([14], p. 177). While Fried concentrates on budgeting risks of death for our ends and values, as well as for the life plan as a whole, at different periods of life, a person's life plan also implies budgets of other sorts of risks as well as of time, energy, and money. Some people may live unreflectively or have incoherent projects, but even they appear to have some ends and values that they are not willing to sacrifice in order to reduce risks to health or to survival. Indeed, a person's 'lifestyle' is determined to a great extent by the ends and values for which he or she will voluntarily take risks. Our willingness to take the risk of ill health or death for success, or friendship, or religious beliefs discloses their importance for us and gives our lives their style. At the very least, governmental intervention in lifestyles and behavioral patterns must meet a very heavy burden of proof because it appears to violate the principle of respect for persons and their self-determination through their life plans and risk budgets.

IV. GOVERNMENTAL INTERVENTION TO REDUCE RISKS

In order to justify governmental intervention into personal lifestyles and behavioral patterns, several conditions should be met. First, justification requires that the intervention have an important end or purpose. This first condition is necessary but not sufficient, for other conditions also have to be met. And several ends and purposes are relevant but not necessarily decisive. They are invoked, for example, in arguments for statutes to require motorcyclists to wear helmets.

Most states have revoked their mandatory helmet statutes in the last few years. But arguments for such statutes have emerged again. Various studies indicate that motorcycle riders without helmets are more likely to suffer fatal head injuries than riders with helmets (see [23] and the studies reported there). Proponents of mandatory helmet legislation offer different reasons for reducing risk-taking (see [26]): (1) Prevention of harm and costs to others or to the society (e.g., motorcycle accidents deprive society of able-bodied people or injured motorcyclists require an excessive amount of society's resources), and (2) Paternalism–prevention of harm and costs to the motorcyclists themselves.

Both reasons invoke beneficence. In the one case, the beneficence is directed primarily at other parties and the society, not at the motorcyclists themselves. In the other case, it is directed primarily at the motorcylists themselves. These reasons can be classified in terms of the primary target of the beneficent policy: the first I will call the *principle of social harm*, the second I will call the *principle of paternalism*. Both rest on or express beneficence. And both may be intermixed as reasons for particular policies.

In order to determine the weight and strength of the reasons, particularly paternalism, I want to examine one defense of self-determination in lifestyles – John Stuart Mill's classic essay "On Liberty" [19]. Mill defended the principle of social harm and rejected the principle of paternalism (at least in its strong form). He asserted what he called "one very simple principle", which is more complex than he realized:

> That principle is that the sole end for which mankind are warranted, individually or collectively, in interfering with the liberty of action of any of their number is self-protection. That the only purpose for which power can be rightfully exercised over any member of a civilized community, against his will, is to prevent harm to others. His own good, either physical or moral, is not a sufficient warrant ([19], p. 68).

Mill's principle depends on a distinction between *self-regarding* and *other-regarding* conduct. This apparently neutral language of 'regard', 'concern', and 'affect' might appear to yield too much to society, for practically all actions regard, affect, and concern others. But it was meant negatively. Mill was interested in conduct that has adverse effects on others. Conduct that concerns, regards, or affects others or the public adversely may be regulated and even prohibited under certain conditions. But conduct that affects only the agent adversely should fall under his/her own self-determination. "In the part which merely concerns himself, his independence is, of right, absolute. Over himself, over his own body and mind, the individual is sovereign" ([19], p. 69). On health, Mill held that "each is the proper guardian of his own health, whether bodily *or* mental or spiritual" ([19], p. 72). Each person is "sovereign" and "final judge" in matters that affect only him or her adversely and is free to conduct "experiments of living" and to make various plans of life ([19], p. 120).

The following chart may illuminate Mill's position.

		Adverse Effects	
		Self-Regarding	Other-Regarding
Voluntariness of action	Voluntary	1	2
	Nonvoluntary	3	4

First, let us examine other-regarding conduct (No. 2 and No. 4). By other-regarding conduct, Mill meant not only conduct that affects others adversely. Such conduct must also affect others directly and without "their free, voluntary, and undeceived consent and participation" ([19], p. 71). If others in the maturity of their faculties consent to the agent's risk-imposition, his action is not other-regarding in Mill's sense. An agent's other-regarding conduct may be either voluntary or nonvoluntary. If it is voluntary (No. 2), if the agent understands the risks and voluntarily imposes them on others without their consent, society may prohibit and punish the conduct through the criminal law. When the agent nonvoluntarily harms others, or imposes risks on them, society may also intervene. Contemporary examples include vaccinations or genetic screening.

When we distinguish voluntary and nonvoluntary conduct, we see that Mill's "simple principle" that self-regarding conduct may not be legitimately restricted or prohibited is complex. He contends that society should not interfere in *voluntary self-regarding conduct*. Such interference would be extreme or strong paternalism. It would override a person's wishes, choices, and actions for his/her own benefit, regardless of his/her condition (e.g., competence). It would be unjustified because it would fail to treat people in the maturity of their faculties as equals; it would be a sign of disrespect.

But interventions in self-regarding conduct that is substantially nonvoluntary because of defects in deciding, willing, and acting may be justified. They are instances of limited or weak paternalism, which does not treat a person with indignity or disrespect. Mill uses the example of a person about to cross a dangerous bridge. If we see such a person, and there is "no time to warn him of his danger," we might "seize him and turn him back, without any real infringement of his liberty; for liberty consists in doing what one desires, and he does not desire to fall into the river" ([19], p. 166). But in this case if the person is not a child, or delirious, etc., only temporary intervention is justified. Once the restrainers determine that the person is acting voluntarily, they must cease and desist. Likewise, Mill argues only for labelling dangerous drugs rather than prohibiting them.

Critics of Mill's principle have tried to take the sting out of it largely by contending that category No. 1 (voluntary self-regarding conduct) is practically a null class because our risky actions are other-regarding and/or nonvoluntary. Thus, they insist, the issue is not whether strong paternalism can be justified but whether most interventions in personal lifeplans and risk budgets are paternalistic at all, at least in the strong sense. To a great extent, the question is where the boundaries of the self and its actions are to be drawn – i.e., whether various 'influences' on the agent 'determine' his actions and thus make them 'nonvoluntary', and whether the adverse effects of his actions extend beyond himself.

V. ADVERSE EFFECTS AND OTHERS

Many critics of Mill contend that few lifestyles and behavioral patterns affect only the agent adversely. Rhetoric abounds: "No man is an island . . . ." "Our society is interdependent." "The distinction between private and public breaks down." Mill did not deny this interdependence ([19], p. 146). He too recognized that most actions affect others, frequently in adverse ways. But harm to others or to the society is only a necessary condition for justified intervention; it may not be sufficient, at least for coercive intervention. Whether the harm is primary or secondary, whether it has the consent of the victim, whether the agent has a definite obligation, and whether the harm outweighs the loss of liberty are all relevant considerations. Mill insisted that an agent's liberty should be overridden only when there is a definite harm, particularly one that violates "a distinct and assignable obligation to any other person or persons . . . " ([19], p. 148). A major question is whether society has a right to claim certain conduct from individuals. Mill was suspicious of attempts to expand the category of harms and obligations to others, particularly in the area of care of one's health. He noted, "If grown persons are to be punished for not taking proper care of themselves, I would rather it were for their own sake than under pretence of preventing them for impairing their capacity or rendering to society benefits which society does not pretend it has a right to exact" ([19], p. 149). The implications are not fully clear, for Mill admitted that each person should bear his/her share of the social burdens, based on some equitable principle, and that society may enforce this obligation through the law ([19], p. 141).

In ill health and early death because of lifestyles, it is often difficult to find direct adverse effects on others. In general, the earlier model of infectious diseases, which justified coercive governmental interventions on grounds of beneficence (and justice), is inapplicable. But some actions such as smoking cigarettes in the work-place and drunken driving do have direct and significant adverse effects on others, and the society may prohibit them if it can also meet other conditions (which will be sketched later). But such prohibitions would need to be carefully circumscribed. Neither smoking nor drunkenness as such should be prohibited; only smoking or drunkenness in settings that result in direct and significant adverse effects on others should be the target of coercive governmental intervention.

Some critics of Mill appear to suppose that it is sufficient to identify some adverse effects on others, however trivial, indirect, or remote. For example, one recent study focuses on 'the energy cost of overweight in the United States.' The authors admit that "overweight (excess body fat) has heretofore been considered mainly a personal problem with important implications for personal health and well-being." But, they contend, "with the rising specters of energy shortage and world hunger, overweight must become a social problem" ([16], p. 765). They establish that overweight is a social problem because of the use of fossil fuel energy to supply the extra food calories to maintain excess body fat.

We have calculated the total fossil energy equivalent of the food calories saved by reducing the present degree of overweight (2.3 billion pounds for the adult United States population) to optimum body weight and the annual fossil energy reduction once all Americans reached their optimum weight. The energy saved by dieting to reach optimal weight is equivalent to 1.3 billion gallons of gasoline and the annual energy savings would more than supply the annual residential electrical demands of Boston, Chicago, San Francisco, and Washington, D. C. ([16], p. 767).

While the authors have established that being overweight involves some moral issues beyond lifestyle, they have not established sufficient grounds for coercive governmental intervention in obesity. The government may, of course, disseminate information and provide services to overweight persons who accept them.

In a complex society, with welfare commitments, it may be possible to emphasize the financial burdens of ill health that certain lifestyles impose on the society and thus on individual taxpayers. These burdens might be construed as harmful (e.g., violation of pecuniary interests) or as unjust (e.g., violation of the principle of fairness in the imposition of burdens). People who engage in risky behavior increase the health care costs of others in private insurance schemes and in Medicare and Medicaid programs. And in a national health insurance program, we can expect widespread opposition to paying for the avoidable afflictions of others.

But more evidence may be needed to establish the costs and the injustice of certain lifestyles and risk budgets. 'Slow suicide' cannot be condemned and prohibited on these grounds without hard evidence that it does in fact impose major costs and unfair burdens on others. Indeed, on a straight cost-benefit analysis, it may be difficult, if not impossible, to criticize those who, in the words of the country-western song, want 'to live fast, love hard, die young, and leave beautiful memories' or who, in the words of the rock song, seek 'life in the fast lane'. Their deaths might actually cost society less than many others. They may die early, without requiring costly care. Thus, a case-by-case analysis is required. For example, even if it is true that motorcycle helmets reduce the number of deaths, we have to consider whether they also lead to survival at a marginal and costly level (e.g., with extensive injuries which require massive medical and social care).

While such considerations may appear to reflect callousness, they are required by an argument for restriction based on the putative costs of certain lifestyles to the society. If these lifestyles are to be restricted because of their alleged costs to society, it is necessary to explore those costs very carefully.

Even if we concede that some lifestyles impose excessive and unfair costs on the society, and that society thus has a right to take some action, it is still necessary to probe the different actions society might take. Prohibition may not be warranted. But if we suppose that obesity and the use of cigarettes and alcohol lead to health problems that require extensive health care, society might institute a tax on these health behaviors to cover the additional medical care that might be required. The risk-takers could pay their own way by bearing the costs of their self-induced medical problems. If their risk-taking is voluntary (an assumption that will be examined later), it is certainly not unfair to have them pay for the costs of these risks. The principle of beneficence does not *require* that society bear the costs of ill health resulting from an individual's voluntary risk-taking. But society's conception of itself as benevolent, even as compassionate and caring, may lead it to provide medical care for the truly needy even if they voluntarily assumed major risks. Furthermore, society may choose to subsidize some forms of risk-taking (such as working in a coal mine, or as a policeman, or as a fireman) because of their importance for society [27].

VI. NONVOLUNTARY ACTIONS

Other critics of the Millian position hold that even if we can identify some *self-regarding* conduct, such conduct is actually nonvoluntary. Thus, they justify weak or limited paternalistic interventions – to protect people from themselves because their choices and actions are substantially nonvoluntary. Resisting all efforts to blame the victims, these critics sometimes defend forms of social determinism that deny almost any individual responsibility for morbidity and early death. For example, Victor Sidel contends:

most health and illness are socially determined rather than individually determined. I refer not only to the obvious instance of environmental pollutants and other unhealthy conditions but also to the fact that most personal health practices are culturally and societally determined ([24], p. 347).

To be sure, cultural and societal forces influence our personal choices such as eating, exercising, smoking, and drinking. But Sidel's denial of personal responsibility is too sweeping. Actions can be *influenced* without being determined, without becoming nonvoluntary or involuntary. As Gerald Dworkin stresses, a person may exercise *second-order autonomy* by affirming those influences or by trying to break them (e.g., by seeking or not seeking help for drug addiction) [10].

It is important to distinguish between direct and indirect paternalism. Most examples involve direct paternalism; for example, a forced transfusion for a Jehovah's Witness. But even though indirect paternalism may not be common, it should not be overlooked. If we ban products such as cigarettes in order to prevent harm to users, we restrict the actions of growers of tobacco and manufacturers of cigarettes. This is indirect paternalism: A is restrained in order to protect B. As stated, indirect paternalism resembles the principle of social harm, but there is one important difference: indirect paternalism concerns harms that require what Gerald Dworkin calls "the active cooperation of the victim" who could avoid the harm if he or she chose to do so ([11], p. 111). Actions that are restricted because of the principle of social harm usually affect individuals against their will (e.g., the manufacture of faulty products).[2]

Of course, self-regarding behavior may be substantially nonvoluntary in some cases. In such cases, weak or limited paternalism may be justified. It is more defensible when it is 'soft' (i.e., when it is directed at helping people realize their own values) than when it is 'hard' (i.e., when it is aimed at imposing alien values). In its 'soft' form, it may, for example, provide information to enable individuals to understand the risks of their actions or try to strengthen weak wills. Its major difficulty, of course, is the determination of the *nonvoluntariness* of lifestyles and behavioral patterns, especially in view of the presumption in favor of acquiescence in personal wishes, choices, and actions when no one else is harmed.

VII. OTHER CONDITIONS FOR JUSTIFIED INTERVENTION

I have concentrated on the ends, based on beneficence, that are relevant to governmental interventions in lifestyles and behavior patterns. But these ends are not sufficient to rebut the presumption against interfering with personal wishes, choices, and actions. Not only do we need strong evidence that the ill health or disease constitutes a net harm or injustice to the society or a net harm to an individual (who is not acting voluntarily); we also need strong evidence that the lifestyle or behavior in question really contributes to ill health or disease.

It is necessary to insist on strong evidence of causal links because terms like 'health' and 'disease' appear to be more objective than they really are. And they can be used to impose value-judgments that are not clearly articulated or defended. For example, when John Knowles, late head of the Rockefeller Foundation, criticized individual lifestyles and behavior patterns, he naturally used the language of traditional vices: "The cost of sloth, gluttony, alcoholic intemperance, reckless driving, sexual frenzy, and smoking is now a national, and not an individual responsibility" ([17], p. 59). Such language may reflect a new moralism. The 'moral police' may operate in the guise of the 'health police' in order to enforce a certain style of life in the name of health when it has little to do with health and more to do with 'morality' and 'salvation'. Indeed, 'moral' and 'religious' beliefs may account for the focus on some actions rather than others; upper middle-class people, for example, are more likely to oppose smoking cigarettes than drinking alcohol.

It is also important to consider the means or mode of governmental intervention. At points in this essay I have focused on coercive policies, in part because they are difficult to justify in terms of beneficence. Occasionally I have suggested that some considerations would justify provision of information but not coercion. Of several possible ways to reduce risk-taking, not all are equally objectionable. For example, provision of information, advice, education, offering incentives, deception, manipulation, behavior modification, and coercion do not equally infringe moral principles such as respect for persons. It is rarely inappropriate to provide information to enable individuals to make their own decisions in a more informed way. And even when stronger measures appear to be justified, whether on grounds of social harm or weak paternalism, the society should choose the least restrictive and least humiliating intervention.

VIII. CONCLUSION

When beneficence is aimed at the reduction of nonvoluntary or involuntary risks of ill health or early death, it offers the strongest reasons for governmental intervention in lifestyles and behavioral patters. This point holds for policies designed to protect people who have not consented to the risks imposed by others (the principle of social harm) and also for policies to protect people who have not voluntarily assumed the risks of their own actions (the principle of weak or limited paternalism). Contemporary debate centers on Mill's category of voluntary self-regarding conduct. Some opponents of Mill's perspective contend that all lifestyles affect others, frequently adversely, or that these lifestyles are substantially nonvoluntary. While we may reject their extreme claims, the difficult task for policy is to determine *which* lifestyles and behavioral patterns actually affect others adversely and *which* are substantially nonvoluntary. In such cases, coercive governmental intervention may be justified, as long as several conditions are met. This debate indicates the importance of empirical data (e.g., regarding the effects and the voluntariness of actions). And it illustrates the impossibility of determining health policy by any single principle such as beneficence.

University of Virginia
Charlottesville, VA

NOTES

[1] Much of this chapter is adapted from James F. Childress, *Who Should Decide? Paternalism in Heath Care* (tentative title) to be published in 1982 by Oxford University Press.

[2] Somewhat similar to indirect paternalism is what may, in fact, be another distinct, nonpaternalistic reason for not acquiescing in a person's risk-taking: the prevention of exploitation. Exploitation may be defined as taking unfair advantage of another person's situation in order to benefit oneself or others. Prevention of exploitation would seem to be an extension of the principle of prevention of social harm or injustice.

BIBLIOGRAPHY

1. Beauchamp, D. E.: 1976, 'Exploring New Ethics for Public Health: Developing a Fair Alcohol Policy', *Journal of Health Politics, Policy and Law* **1**, 338–354.
2. Beauchamp, D. E.: 1980, 'Public Health and Individual Liberty', *Annual Review of Public Health* **1**, 121–136.
3. Beauchamp, D. E.: 1976, 'Public Health as Social Justice', *Inquiry* **13**, 3–14.
4. Beauchamp, T. L. and Childress, J. F.: 1979, *Principles of Biomedical Ethics*, Oxford University Press, New York.
5. Belloc, N. B.: 1973, 'Relationship of Health Practices and Mortality', *Preventive Medicine* **2**, 67–81.

6. Belloc, N. B. and Breslow, L.: 1972, 'Relationship of Physical Status and Health Practices', *Preventive Medicine* **1**, 409–421.
7. Brailey, A. G., Jr.: 1980, 'The Promotion of Health through Health Insurance', *The New England Journal of Medicine* **302**, 51–52.
8. Childress, J. F.: 1981, 'Priorities in the Allocation of Health Care Resources', in E. E. Shelp (ed.), *Justice and Health Care*, D. Reidel Publ. Co., Dordrecht, Holland; Boston, U.S.A., pp. 139–150.
9. Childress, J. F.: 1980, 'Risk', *The Encyclopedia of Bioethics*, MacMillan-Free Press, New York, pp. 1516 –1522.
10. Dworkin, G.: 1976, 'Autonomy and Behavior Control', *The Hastings Center Report* **6** (February), 23–28.
11. Dworkin, G.: 1971, 'Paternalism', in R. A. Wasserstrom (ed.), *Morality and the Law*, Wadsworth Publ. Co., Belmont, California, pp. 107 – 126.
12. Frankena, W. K.: 1973, *Ethics*, 2nd ed., Prentice-Hall, Inc., Englewood Cliffs, N.J.
13. Freymann, J. G.: 1975, 'Medicine's Great Schism: Prevention vs. Cure: An Historical Interpretation', *Medical Care* **12**, 525–536.
14. Fried, C.: 1970, *An Anatomy of Values: Problems of Personal and Social Choice*, Harvard University Press, Cambridge.
15. Fuchs, V. R.: 1974, *Who Shall Live? Health, Economics, and Social Choice*, Basic Books, New York.
16. Hannon, B. M. and Lohman, T. G.: 1978, 'The Energy Cost of Overweight in the United States', *American Journal of Public Health* **68**, 765–767.
17. Knowles, J. H.: 1977, 'The Responsibility of the Individual', in J. H. Knowles (ed.), *Doing Better and Feeling Worse: Health in the United States*, W. W. Norton, New York, pp. 57–80.
18. Lurie, A.: 1975, *The War Between the Tates*, Warner Books, New York.
19. Mill, J. S.: 1976, *On Liberty*, G. Himmelfarb (ed.), Penguin Books, Harmondsworth.
20. National Commission for the Protection of Human Subjects: 1978, *The Belmont Report: Ethical Principles and Guidelines for the Protection of Human Subjects of Research*, DHEW No. (OS) 78–0012, Washington, D. C.
21. National Heart, Lung and Blood Institute: 1979, *Proceedings of the Conference on the Decline in Coronary Heart Disease Mortality*, NIH Publication No. 79–1610.
22. Nightingale, E. O. *et al.*: 1978, *Perspectives on Health Promotion and Disease Prevention in the United States*, A Staff Paper of the Institute of Medicine, National Academy of Sciences, Washington, D. C.
23. Russo, P. K.: 1978, 'Easy Rider – Hard Facts: Motorcycle Helmet Laws', *The New England Journal of Medicine* **299**, 1074–1076.
24. Sidel, V. W.: 1978, 'The Right to Health Care: An International Perspective,' in E. L. Bandman and B. Bandman (eds.), *Bioethics and Human Rights: A Reader for Health Professionals*, Little, Brown and Company, Boston, pp. 341–350.
25. Starr, C.: 1969, 'Social Benefit versus Technological Risk', *Science* **165**, 1232–1238.
26. Tribe, L. H.: 1978, *American Constitutional Law*, The Foundation Press, Mineola, N. Y.
27. Veatch, R. M.: 1980, 'Voluntary Risks to Health: The Ethical Issues', *Journal of the American Medical Association* **243**, 50–55.

28. Warmington, E. H. and Rouse, P. G. (eds.): 1956, *Great Dialogues of Plato*, transl. W. H. D. Rouse, The New American Library, New York.
29. Walker, W. J.: 1977, 'Changing United States Life-style and Declining Vascular Mortality: Cause or Coincidence?', *The New England Journal of Medicine* **297**, 163–165.
30. Wikler, D. I.: 1978, 'Persuasion and Coercion for Health: Ethical Issues in Government Efforts to Change Life-Styles', *Milbank Memorial Fund Quarterly/Health and Society* **56**, 303–338.

RONALD M. GREEN

ALTRUISM IN HEALTH CARE

Can a commitment to social justice and to altruism in medicine coexist? Can medicine remain a locus of human generosity and compassion as health care provision becomes more and more a social responsibility? Over the past decade a number of writers, myself among them, have sought to assert that there is a fundamental human right to health care. In different ways we have argued that access to health care cannot be allowed to depend upon one's financial resources or on the free generosity of some health care providers. Each person, we have said, has a legitimate moral *claim* to a fair share of the health resources a society provides ([11, 12, 24, 33]). But even as we have argued this way, it has been incumbent upon us to consider the *costs* of the right we would affirm. On the economic side, this matter has received attention as thinkers have explored the problem of how quality of care can be conjoined with equality of its distribution ([4, 9]). But far less attention has been given to the possible moral costs of this right. In what follows I want to examine one of these possible costs by asking whether some of the altruistic moral attitudes and practices traditionally associated with medical care can survive in an environment of social or legally established rights and claims.

The question and problem I have in mind become clearer if we recall that altruism and health care have historically had a very close relationship to one another. Altruism, which we might loosely define as spontaneous and undemanded devotion to other persons' interests and welfare, has long found very special expression in the medical setting. In turn, the medical setting has frequently been regarded as a particularly suitable environment for the development of altruism and for the stimulation of attitudes of generosity and compassion. The first of these relationships, the dependence of health care on altruism, is evidenced throughout the history of medical institutions and medical practice. For example, the foundation of hospitals in the West was largely stimulated by the spirit of Christian charity. These institutions usually had their beginnings in the hospices, almshouses and sick bays of early monastic institutions ([10], pp. 176–189). Up through the Middle Ages, in fact, hospitals were staffed not by doctors but by monks and nuns whose service was viewed as a consummate expression of Christian charity.

Earl E. Shelp (ed.), Beneficence and Health Care, 239–254.

In the modern period the development of the physician's role and professional responsibilities helped shift the burden of care to a medical team, but the view of this care as an expression of undemanded and undemandable altruism remained substantially intact. It is true that many of the treatises on medical ethics and professional codes that emerge during this period stress the physician's obligations to his patients and his responsibility to assist in the care of the poor or unfortunate. But these duties are not seen as based on *rights* possessed by patients so much as on the personal sense of duty and benevolence possessed by the virtuous physician ([21], p. 207; [22], pp. 63ff). Prior to the very modern period it is only in revolutionary France that we encounter a conception of access to health care grounded not in the physician's integrity or generosity but in the citizen's right to medical treatment ([22], p. 57).

The traditional association of altruism and health care finds expression not only in the view that philanthropic motives should underlie care for the sick, but also in the reciprocal notion that the medical setting provides a unique and valuable opportunity for the development of human compassion. This understanding of medicine and medical care as a kind of moral pedagogy has deep roots in the Western religious traditions where visitation of the sick, nursing and healing, in addition to their direct worth for the patient, are commonly valued as devotional exercises and opportunities for personal spiritual and moral growth ([2], pp. 938–942; [13], pp. 106–111). But a similar view is sometimes found outside of the religious communities in the secular tradition of moral philosophy. Immanuel Kant, for example, saw great moral value in attendance on the sick. Although Kant insisted that the duties of love and benevolence involved forms of conduct, and not emotional states (since such states could never be commanded), he nevertheless believed that we are required to try to cultivate sympathetic natural feelings as an incentive to duty. For this reason, he maintained, human beings have an obligation not to avoid the reality of suffering but to seek it out in order to render assistance. In this connection he expressly advocated regular visitation of the sick ([16], p. 124).

It is quite easy to appreciate this constant and reciprocal linking of health care and altruism in our cultural past. In many ways it seems entirely fitting that the sick be in the hands of individuals substantially motivated by a sense of human compassion. Nevertheless, in our own day a number of forces have joined to weaken this traditional link. Within the social sphere, for example, the increasing pace of medical practice, the concentration of care in larger and less personal institutions and the growing commercialization of medical services have, in many persons' view, contributed to an erosion of older habits of compassion. The intensive technologizing of medicine has also helped to weaken personal relationships in the medical setting and, in some cases, has interposed real physical barriers between the patient and care givers or friends. Thus, the modern intensive care unit surely does not facilitate the traditional rites of visitation or much display of compassion. But there are more than social or technological reasons for a possibly declining place for direct expressions of altruism in medical care. Moral considerations may also have contributed to this development. These can be summed up in the increasingly prevalent belief that health care should not be thought of as something freely given by one generous group of persons to others in need, but rather as a fully legitimate claim by persons upon society and the community of health care providers. Behind this changing conception of health care, moreover, lie several specific moral considerations that have undermined the traditional value placed on altruism in the medical setting. Paradoxically, the strong traditional idea that medicine should display the purest and most intensive instances of human altruism may have contributed to the reduced estimate of personal acts of compassion or generosity in this domain.

One moral problem with personal acts of charity or compassion is quite familiar and is not confined to medical altruism. It is the problem that philanthropy in the broadest sense can often be a means for the subtle exercise of control and domination. As Reinhold Niebuhr has noted, behind a benevolent facade, the philanthropist may often be a brutal "man of power" who chooses to dispense favors at his discretion in order to bind people to himself and to forestall the fair social distribution of resources. This philanthropist's generosity, Niebuhr observes, "is at once a display of his power and an expression of his pity" ([23], p. 14). In a similar vein, Schwartz has noted that individuals engaged in public generosity to others frequently employ a kind of "gratitude imperative" as a tool for the protection of their status and control ([27], p. 73).

In our own day, these moral flaws in private philanthropy often have been found in forms of medical altruism. For example, what previous generations often regarded favorably as compassion on the part of physicians has come to be viewed more as a form of self-serving paternalism. Surveying medical codes since the nineteenth century, William F. May finds support for this harsh judgment in the tones of condescension the codes display and in what he calls their "conceit of philanthropy". Rather than properly viewing the physician's commitment to patients as an outgrowth of his location in a community and his inescapable indebtedness to others, says May, these codes typically regard the physician's acts as wholly gratuitious. The result is a self-serving and erroneous picture of a "self-sufficient monad who out of the nobility and generosity of his disposition and the gratuitiously accepted conscience of his profession has taken upon himself the noble life of service" ([19], pp. 70f). May's negative estimate here is clearly widely shared. Acts and forms of bearing among medical professionals once regarded with awe and respect have in our era come to be viewed with deep moral suspicion.

In the contemporary period, therefore, medical care systems and medical professionals have experienced something of a dilemma. The generous and, to a degree, spontaneous provision of medical care probably remains as much an exemplification of human giving as ever, yet many traditional personal displays of such altruism have been rendered morally suspect. How, then, can these seemingly conflicting moral sentiments be harmonized? How can altruism be preserved, fostered and yet freed from the possibility of personal abuse? Very typically, an answer to this question has been found in institutional and social procedures which separate the altruist from the recipient of generosity. In some cases this separation is directly imposed on the process of giving or helping. During the early days of kidney transplantation, for example, it was often observed that a kind of creditor-debtor relationship developed between organ donors and recipients that proved emotionally disabling to both partners. Indeed, even when cadaver donors were involved, organ recipients often developed a kind of crippling emotional indebtedness to the person whose death had made their own survival possible ([8], p. 383). In order to reduce this 'tyranny of the gift' it has become standard practice for members of the medical team to serve as gate-keepers between donor and recipient. In the case of cadaver organs, efforts are made to preserve the anonymity of the donor.

On a grander scale, this same logic has been applied to the provision of health care generally. Morally uncomfortable with an approach that makes care dependent on a series of personal acts of generosity, modern societies have gravitated towards an alternate system based on what some have called 'political' or 'institutional' altruism ([25], p. 173; [31], p. 5). This involves the provision of health care through tax-supported governmental schemes. Many other motivating factors, of course, beyond a moral discomfort with private generosity have been at work in the establishment of national health services or national health insurance programs. Economists observe that collective programs like these can be viewed as a form of self-protective 'sickness insurance', a means by which individuals make provision for their own future health care needs [5]. These concerns have been exacerbated by the rising cost of modern medical care and by the breakdown or inadequacy of the older 'charity market' means of provision to those in need of treatment [14]. Others see these programs as means by which the modern state has endeavored to protect itself against the dependency burden created by illness among the poor ([6], p. 76). And even less favorably, some regard these programs as involving a kind of 'Robin Hood' altruism or 'moralistic aggression' whereby poorer voters use collective power to transfer resources from the more prosperous ([31], p. 42).

While these or other motivations may partly explain the wide appeal of the state support of medical services in our day, we miss the full moral significance of this development if we fail to see its continuity and discontinuity with older moral conceptions of medical care. In fact, an older appreciation of medical altruism seems to have combined here with a discomfort over possible abuses of altruism to stimulate the development of state intervention in this area. Public attitudes in Great Britain toward the National Health Service and public debate in this country over the Medicaid and Medicare programs evidence a profound interweaving of altruistic concern for the indigent infirm and a preoccupation with the insurance or transfer aspects of these programs.

On a scholarly level, these complex contemporary altruistic considerations have been given forceful expression by Richard Titmuss in his now classic book, *The Gift Relationship* [32]. Ostensibly, Titmuss's study aims at a comparative assessment of the British and American systems for the collection and distribution of human blood. In this assessment the American system with its substantial reliance on paid as against voluntary donors emerges as inferior. Some problems identified by Titmuss have to do with the medical dangers and relative inefficiency of the American system. Because of its reliance on donors for whom money is a major inducement, this system attracts less healthy contributors or those prone to deception about their health. The blood collected can thus be contaminated and, for this and other reasons, is more subject to being wasted. But these technical and medical considerations are not at the heart of Titmuss's preference for the voluntary British system. Rather, the commercial exchange of blood in his view has a serious moral deficiency because it interferes with the expression of generosity between persons and because it places the burden of social giving on those least able to afford it. In Titmuss's own words:

From our study of the private market in blood in the United States we have concluded that the commercialization of blood and donor relationships represses the expression of altruism, erodes the sense of community, lowers scientific standards, limits both personal and professional freedoms, sanctions the making of profits in hospitals and clinical laboratories, legalizes hostility between doctor and patient, subjects critical areas of medicine to the laws of the marketplace, places immense social costs on those least able to bear them – the poor, the sick and inept – increases the danger of unethical behavior in various sectors of medical science and practice, and results in situations where proportionately more and more blood is supplied by the poor, the unskilled, the unemployed, Negroes and other low income groups and categories of exploited human populations of high blood yielders ([32], pp. 245f).

I have suggested – and I believe this suggestion is borne out by this last remark – that Titmuss's concern extends beyond the technical details of blood collection systems to the deeper moral implications of these systems. But I would also contend that Titmuss's interests go beyond the matter of blood collection per se and embrace the much larger issue of medical altruism in the modern setting. What Titmuss wishes to defend is the British National Health Service itself, with its emphasis on the collective social provision of health care. In contrast to the American system of fee-for-service medicine and proprietary medical institutions, Titmuss appears to believe that the British health system, no less than its blood collection program, fosters the essential altruism or 'gift relationship' which is vital to the moral health of a society.

It is true that Titmuss does not fully develop this larger theme, and it is also true that in some ways there is an important moral difference between the NHS and the blood program that is the specific object of his study. The health system involves less immediate spontaneity of giving based as it is on a coercive system of taxation. Nevertheless, there are also important similarities between these two. To some degree, both elicit a measure of altruistic commitment to others, in one case when a decision to donate blood is made and in the other, perhaps in the privacy of the voting booth, when the health system is established and its funding levels set. Then, too, – and in Titmuss's view this may really be central – both involve a willingness to care for 'unnamed strangers'. It is the essential anonymity and impersonality of these forms of medical altruism that seems to underlie their powerful moral appeal to Titmuss and others who share his view. In a sense, anonymity functions as a guarantee of the relative purity of the altruism that lies behind these systems of care, a testimony to the truly spontaneous expression of social feeling they represent. The anonymity of recipients carries compassion beyond the socially limited sphere of one's immediate family and friends to society as a whole. And the fact that the distant recipients of one's compassion are unable to reward these acts or personally to express their gratitude frees altruism from any hint, however subtle, of selfishness, coercion or control.

As one might imagine, Titmuss's discussion has generated a great deal of controversy. Many of his statistical and medical claims have been challenged in the effort to defend the adequacy of commercial blood services and (by implication) the free market provision of health care ([7, 25]). None of these criticisms, however, go to the heart of Titmuss's moral vision and in many ways this is the most interesting aspect of his work. If what I have been saying is correct, Titmuss's argument illustrates the changing form which the conception of medical altruism has taken in our day. This newer conception is in continuity with the old: it embodies a deep commitment to the provision of medical care for those who need it – especially the indigent and others whose circumstances compound their physical suffering. But in contrast to older ideas, this provision is ideally held to be impersonal, a commitment to 'unnamed strangers', and it is mediated through laws, social institutions or procedures which separate those motivated to help from the recipients of their generosity and concern. In this respect, the newer conception of medical altruism rests on something of a paradox: intense social feeling and interpersonal concern is channeled through impersonal institutions. In a criticism of Titmuss's view, Kenneth Arrow notes the peculiarity of this position:

> Indeed there is something of a paradox in Titmuss' philosophy. He is especially interested in the expression of impersonal altruism. It is not the richness of family relationships or the close ties of a small community which he wishes to promote. It is rather a diffuse expression of confidence by individuals in the workings of society as a whole. But such an expression of impersonal altruism is as far removed from the feelings of personal interaction as any marketplace ([3], p. 5).

No less than other economic critics, I think, Arrow misses the deeper moral vision in Titmuss's work. But his critical observation here is worth bearing in mind. And it is especially relevant to those persons who would follow Titmuss in advocating essentially impersonal forms of medical altruism. To an extent, those of us who have argued for a basic human right to health care are in this class. For although the essence of these arguments is that health care is something due persons and is not to be thought of as a free gift from others – hence the stress on justice and rights in this context – the framers of these arguments and the advocates of a right to health care are surely motivated by a commitment to persons' dignity and well-being and by a moral vision of social solidarity similar to that possessed by Titmuss. Indeed, the stimulus behind many modern social welfare programs, not just those in health care, is an older kind of altruism combined with a contemporary moral sense that there should be, as one writer puts it, "a degree of social distance between helped and helper" ([34], p. 141). Concepts of justice and rights partly express and protect this necessary distance. But if this is so, appeals for a 'right' to health care and systems founded on that right display much of the same paradox as that found in Titmuss's work: a deliberately impersonal series of ideas and arrangements is made the vehicle of deep personal concerns.

Superficially this paradox may not appear to present any problems. Modern welfare programs have often functioned effectively to meet needs once met by private charity and philanthropy. Some trade-offs or losses on the personal level may be imposed by this transfer to governmental authority but these have usually been considered an acceptable price to pay for the moral and material superiority of this approach. Nevertheless, those who advocate the increasingly social assumption of a responsibility for providing medical care, based on a right to health care, must consider and anticipate some of the moral costs of this transformation. Precisely because medical care systems have so traditionally been characterized throughout by interpersonal expressions of compassion, transition to less personal modes of service founded on claims of human rights may pose new kinds of moral problems or challenges. In a rather cursory fashion I want now to suggest some ways in which the widespread establishment and effectuation of a right to health care might endanger the expression of altruism in the medical setting. None of these possible dangers justifies qualifying the idea of a right to health care, I think. But they do require some forethought if we allow ourselves to move progressively toward a system founded on more impersonal notions of social sharing.

One problem we might anticipate as a right of equal access to health care becomes established is already familiar to us in the domain of social welfare generally. For want of a better word we might call it the 'backlash' problem. It derives from limits on the resources of altruism on which a society may draw. The problem is illustrated by the complaint voiced by a taxpayer forced to wait in long lines in a hospital clinic while many Medicaid patients are attended before her [20]. Her bitterness on being told by a clinic employee that her private insurance forms are more of a nuisance to process than the customary Medicaid forms and her resentment at the deference and seeming preference given to what she regards as 'charity' patients highlight the difficulties facing impersonal altruism in complex societies. Where a socially protected right to health care is concerned, these difficulties are compounded by many factors. The 'insurance' aspect of these programs obscures for many persons their altruistic dimensions and does not prepare individuals to view the moral significance of widely distributing medical care. Then, too, the transfer implications of these programs naturally meet resistance from those who dislike the provision of specific forms or degrees of medical support. It can be argued that the bitterness and resentment of this backlash phenomenon evidences a profound misunderstanding of the sense in which equal access to health care is not a gift from some persons to others, but a basic human right. And this makes all the more important the task before those who would advocate this right or help put it into effect. Nevertheless, we must not forget that while medicine has always been a central setting for expressing altruism, these expressions usually have been personal and self-selected by caregivers. As society moves to the less personal forms of sharing, as it deliberately institutionalizes the distance between 'helped and helper' it must anticipate deep strains and tensions.

Similar strains and tensions might be expected in the specific relationship between health care professionals and patients. We have seen that traditionally this relationship was based on compassion and, to a degree, on condescension in both the best and the worst senses of this word. The physician (and the nurse) was regarded as altruistically assuming responsibilities and, in many instances also, as generously dispensing his or her time, skill and emotional concern. But what happens to this relationship when the patient is viewed as having legitimate claims on these resources? Earlier I mentioned William May's criticism of the purely philanthropic understanding of the physician's role. Despite these criticisms, however, May also laments the possible notion of a 'contractualist' model of medical care where the physician or other members of the health care team primarily regard themselves as legally or financially committed to render service because of contractual relationships into which they have entered. May contrasts this 'contractual' model with a more organic 'convenantal' vision of the physician-patient relationship. He believes the contractual model holds great dangers for medical care since it encourages a kind of 'minimalism' of service in which the professional becomes "too grudging, too calculating, too lacking in spontaneity, too quickly exhausted to go the second mile with his patients along the road of their distress" ([19], p. 73). The question, of course, is whether a health care system embodying a right of all citizens to equal access to good quality care, a system in which physicians are viewed as providers of care to socially designated recipients, will generate 'contractualist' medicine of this sort.

In a recent article, a physician, Mark Siegler, has argued that the establishment of a right to health care will inevitably lead to replacement of the older 'covenantal' and compassionate model of care with the kind of contractualist model May decries. Siegler believes that along with other considerations this constitutes a good reason for rejecting the idea of a right to health care in the first place:

> The traditional covenantal relationship between doctor and patient goes well beyond contractual minimalism and maximalism. It includes an element of gift that is characteristic of most covenants. The notion of covenant involves a loyalty and fidelity to promise beyond that of commercial transactions. Contract presupposes an argeement based on mutual self-interest; covenantal ethics encompasses concepts of donation but also contains elements of human indebtedness and human responsibility. I remain convinced that the granting of a right to healtth care would lead to the establishment of a contractual model of medicine, and that this would diminish the quality and effectiveness of the traditional doctor-patient relationship ([30], pp. 154f).

A related criticism has also been voiced by another physician, Eric Cassell, in an article very appropriately entitled "Do Justice, Love Mercy: The Inappropriateness of the Concept of Justice Applied to Bedside Decisions" [6]. Cassell fears that with the growing concept of justice in health care the compassionate physician will be transformed into an impersonal agent of society. He especially laments the idea that the physician may be called upon to make or to carry out difficult priority and allocative decisions imposed by social constraints on the provision of health care. The logo of the New York Hospital, Cassell reminds us, is not blindfolded justice, but the Good Samaritan ([6], p. 79).

In different ways, then, these writers are drawing our attention to the possible moral cost a right to health care may impose on one of the most cherished and most altruistic relationships our society possesses. Their concerns are real ones. At the same time, these arguments against a right to health care betray certain confusions and unclarities. Surely the depersonalization of the care situation does not result only from appeals or steps toward social justice in this area. The very tendency toward 'contractual' medicine May criticizes, for example, has largely arisen on the soil of fee-for-service and propriety medical care. Altruism, as Titmuss sees, can be threatened as much by commercialism as by other forces. In this respect the establishment of more just systems of health care holds at least as much promise for restoring compassion to the medical setting as it does threat.

Beyond this, the quality of interpersonal concern in a health care system is a complex matter. It results as much from the education and the self-understanding of professionals as it does from surrounding institutional frameworks or society's expectations. Cassell himself traces much of the depersonalization of medicine today to the processes of medical education. Even before their interviews at medical school, he observes, prospective students learn that it sounds too sentimental to explain one's commitment to medicine in terms of wanting to 'help people.' Subsequent years of intensive scientific training obliterate this motive even more, so that by the time they graduate, students "may not even remember, not to mention publicly acknowledge, why they chose medicine in the first place" ([6], p. 80). Against this background, it is far too simple to relate fears of declining altruism or human compassion in the medical profession to appeals for justice. What these fears indicate, however, is the need for attention to maintaining or renewing traditional habits of compassion on the part of medical professionals even as society moves toward a more institutionalized and just provision of health care. These goals are not necessarily exclusive and, to some extent, they may even be complementary. But traditional professional attitudes of altruism will not endure in a changing moral and social environment unless these traditional attitudes are valued and cultivated. Also important here, if Cassell's worries are taken seriously, is careful reflection on mechanisms for insulating medical professionals from some of the difficult economic and allocative decisions that establishment of an equal right to high quality care may bring in its train.

A final peril to medical altruism in our day may also be only tangentially related to the establishment of a right to health care, but the association is firm enough in some people's minds that the matter requires attention. I refer to the impact of this right on private acts of altruism in the areas of medical research and organ donation. In fact, several problems are involved here. Where research is concerned, for example, equal access to health care is seen as possibly reducing the supply of traditional volunteers for experimental studies and therapies. At least until very recently, the indigent furnished a pool of experimental subjects in large teaching hospitals and research institutions. With the transition to a system of more equal access to quality health care, however, some see this pool as drying up and they wonder whether equal access to care should not now be thought of as implying equal susceptibility to participation in medical research. Thomas Almy expresses this point of view very well:

> The process of continuous growth and renewal of biomedical knowledge, as well as the cultivation of professional skills, cannot benefit the next generation without the help of the patients of today. We have resolved that this burden shall no longer be borne chiefly by the poor, the ignorant and wards of the state. If we take an egalitarian view of the *rights* of the patient, should we take any other view of his *responsibilities*? ([1], p. 166).

Almy's question here is meant to be provocative and he does not appear personally to advocate compulsory involvement in research as a condition of equalized access to medical care. But the logic behind proposals for mandatory participation in research is, to a certain extent, compelling, and we should not be surprised if, as the supply of present research subjects dwindles, some persons advocate putting more coercive ideas into practice.

The issue of organ transplantation also raises questions of obligation and compulsory social service in the setting of equal access. The problem here is less a dwindling supply of voluntary donations than a burgeoning demand for both live and cadaver organs. This increase in demand, of course, is largely a result of the increased success rate of various forms of transplant surgery. But the availability of state funding for complex and prolonged medical procedures like this has surely played a part in stimulating the demand for organs as well. In addition, the development of state support for costly technologies like renal dialysis has greatly expanded the population eligible for organ transplants.

Against this background, it is natural to hear appeals for a more expeditious way of securing organs than to await express donation by individuals. While no one yet advocates compulsory donation of one of paired organs or of replenishable bodily substances like bone marrow on the part of living donors, there have been periodic calls for what Paul Ramsey has called the "routine salvaging of cadaver organs" ([26], p. 198). Recently, debate has surfaced in Great Britain over legislative proposals to replace the present 'opting in' system with an 'opting out' system of organ donations. Under these proposals, an individual's organs would become available for transplant at his death unless he had expressly refused consent while alive or unless his close relatives expressly objected ([17, 28]). Thus an act that has usually in this country and Great Britain been an expression of personal generosity would by some persons be converted into an expected fulfillment of social responsibility.

What can one say about all of these proposals for mandating these special forms of medical altruism? Are they a natural and unavoidable extension of the logic of impersonal altruism that partly underlies the idea of a right to health care? Does a required provision of medical services necessitate required giving through participation in research or organ donation on the part of individuals?

Let me simply say that I hope the answer to these questions is no. With Titmuss I share the view that genuine and spontaneous altruism, whether to a named or unnamed other, is a precious social resource. Its expression should in every way be encouraged, and this is especially true in the medical setting where basic matters of life and death are involved and where our common humanity and mutual need are so much on display. Unless there is no other option or unless impersonal and mandated giving is the only alternative to abusive or exploitive private charity, opportunities for the personal expression of altruism should always be welcomed and encouraged.

As we consider these various proposals for compulsion, it seems reasonably clear that social needs in the areas of research or transplantation are nowhere near the point where the replacement of altruism and generosity by obligation and requirement is either necessary or desirable. Indeed, these are both areas where voluntary participation, even when it is a less efficient way of meeting social needs, is something intrinsically good and worth preserving. Scientific research, for example, has drawn upon a long tradition of volunteer participation which has value for those involved as subjects and, through their example, for society as a whole. Hans Jonas describes this tradition of idealistic generosity as a kind of social 'capital' which renews itself and grows only if the essential voluntariness of participation is preserved:

> Freedom is certainly the first condition to be observed here. The surrender of one's body to medical experimentation is entirely outside the enforceable 'social contract.' ... Indeed, we must look outside the sphere of the social contract, outside the whole realm of public rights and duties, for the motivation and norms by which we can expect ever again the upwelling of a will to give what nobody – neither society nor fellow man, nor posterity – is entitled to ([15], p. 309).

Jonas qualifies this insistence on freedom by acknowledging that in times of grave crisis society may tolerate compulsion in this area and may risk its moral 'capital' of free idealism. But he sees this as permissible only when needed in moments of great urgency – situations of 'clear and present danger'. Outside these circumstances, to compel persons to serve as research subjects is unacceptable both because it violates human personhood and because it jeopardizes the deeper resources of social altruism.

Very much the same reasoning would seem to apply to organ donation. Whether involving a living or deceased donor, the bestowal of an organ, as Fox and Swazey note, represents an 'ultimate' expression of the Jewish and Christian ethic of self-sacrificial giving to others ([8], p. 381). We have seen that an element of impersonality in the process of donation may sometimes be required in order psychologically to protect either the donor or the recipient. But certainly the need for impersonality here need not extend to institutionalized compulsion or requirement. So long as an adequate or equivalent supply of organs can be procured by customary means, every effort should be made to preserve the possibility of genuine 'donation', of what one writer describes as "the interpersonal character of transplantation and the aspect of humane and voluntary service of another which is intrinsic to it" ([18], p. 69). This is so, once again, because spontaneous giving is a value not only to the individual donor but to society as a whole. Acts of this sort nourish the social spirit and through their example they help soften the contours of a society otherwise governed by sterner legal and moral requirements.

Oddly enough, as we move toward a more common acceptance of the idea of a right to health care, there may even be an increasing need for opportunities for personal expression of altruism of this sort. As we have seen, medical care has always provided occasions for the exercise of compassion and generosity. We also have seen that it is no contradiction to this that in our own day the older spirit of medical altruism finds partial expression in appeals for the social provision of health care and the acknowledgement of a 'right' to health care. Nevertheless, there are inherent tensions between these two facets of our moral tradition, tensions that may be relieved if within the context of the social provision of health care we make as much room as we possibly can for generosity and compassion between persons. This means that we must resist the logic that would convert social rights into an unlimited array of corresponding social responsibilities, with no room left for free giving.

It would appear that a health care system without justice is no longer acceptable to many of us. But, as Earl Shelp reminds us, "[t]o limit the moral test of health care and health policy to justice runs the risk of limiting morality to duty at the expense of other desirable attributes such as beneficence, sacrifice, and compassion . . . " ([29], p. 226). A truly just and praiseworthy health care system, therefore, must embrace these other values as well. It must have room for and it must help create genuinely altruistic professionals committed, beyond 'minimalism,' to human service. It must seek to multiply occasions for free human giving and for the expression of compassion between persons. Above all it must never allow the routinization and institutionalization that may accompany a right to health care to efface the genuine concern for persons that underlies the insistence on that right in the first place.

Dartmouth College
Hanover, New Hampshire

BIBLIOGRAPHY

1. Almy, T.: 1977, 'Meditation on a Forest Path', *New England Journal of Medicine* **297**, 165–167.
2. Amundsen, D. W.: 1978, 'History of Medical Ethics, Medieval Europe', in W. Reich (ed.), *Encyclopedia of Bioethics*, Vol. 2, Macmillan Co., Free Press, New York, pp. 938–950.
3. Arrow, K.: 1972, 'Gifts and Exchanges', *Philosophy and Public Affairs* **1**, 343–362.
4. Arrow, K.: 1973, 'Some Ordinalist-Utilitarian Notes on Rawls's Theory of Justice', *Journal of Philosophy* **70**, 239–257.
5. Buchanan, J. R. and Tullock, G.: 1965, *The Calculus of Consent*, University of Michigan Press, Ann Arbor.
6. Cassell, E. J.: 1981, 'Do Justice, Love Mercy: The Inappropriateness of the Concept of Justice Applied to Bedside Decisions', in E. E. Shelp (ed.), *Justice and Health Care*, D. Reidel Publ. Co., Dordrecht, Holland; Boston, U.S.A., pp. 75–82.
7. Cooper, M. C. and Culyer, A. J.: 1973, 'The Economics of Giving and Selling Blood', in *The Economics of Charity*, Institute of Economic Affairs, London, pp. 79–101.
8. Fox, R. C. and Swazey, J. P.: 1978, *The Courage to Fail*, 2nd ed., University of Chicago Press, Chicago.
9. Fried, C.: 1976, 'Equality and Rights in Medical Care', in J. G. Perpich (ed.), *Implications of Guaranteeing Medical Care*, National Academy of Sciences, Institute of Medicine, Washington, D.C., pp. 3–14.
10. Garrison, F. H.: 1966, *An Introduction to the History of Medicine*, 4th ed., W. B. Saunders Co., Philadelphia.
11. Green, R. M.: 1976, 'Health Care and Justice in Contract Theory Perspective', in R. M. Veatch and R. Branson (eds.), *Ethics and Health Policy*, Ballinger Publ. Co., Cambridge, Mass., pp. 111–126.
12. Green, R. M.: 1981, 'Justice and the Claims of Future Generations', in E. E. Shelp (ed.), *Justice and Health Care*, D. Reidel Publ. Co., Dordrecht, Holland; Boston, U.S.A., pp. 193–211.
13. Jakobovits, I.: 1959, *Jewish Medical Ethics*, Bloch Publ. Co., New York.
14. Johnson, D. B.: 1973, 'The Charity Market: Theory and Practice', in *The Economics of Charity*, The Institute of Economic Affairs, London, pp. 79–101.
15. Jonas, H.: 1977, 'Philosophical Reflections on Experimenting with Human Subjects', in S. J. Reiser, *et al*. (eds.), *Ethics in Medicine*, M.I.T. Press, Cambridge, Mass., pp. 304–315.
16. Kant, I.: 1964, *The Doctrine of Virtue, Part II of the Metaphysics of Morals*, University of Pennsylvania Press, Philadelphia.
17. Kennedy, I.: 1979, 'The Donation and Transplantation of Kidneys: Should the Law be Changed?', *Journal of Medical Ethics* **5**, 13–21.
18. Mahoney, J.: 1975, 'Ethical Aspects of Donor Consent in Transplantation', *Journal of Medical Ethics* **1**, 67–70.
19. May, W. F.: 1977, 'Code and Covenant or Philanthropy and Contract', in S. J. Reiser *et al*. (eds.), *Ethics in Medicine*, M.I.T. Press, Cambridge, Mass., pp. 56–76.
20. McCrae, S.: 1981, 'When They Make You Wait', *New York Times*, April 27, p. 26.

21. McCullough, L. B.: 1979, 'Rights, Health Care and Public Policy', *Journal of Medicine and Philosophy* 4, 204–215.
22. McCullough, L. B.: 1981, 'Justice and Health Care: Historical Perspectives and Precedents', in E. E. Shelp (ed.), *Justice and Health Care*, D. Reidel Publ. Co., Dordrecht, Holland; Boston, U.S.A., pp. 51–71.
23. Niebuhr, R.: 1932, *Moral Man and Immoral Society*, Charles Scribner's Sons, New York.
24. Outka, G.: 1974, 'Social Justice and Equal Access to Health Care', *Journal of Religious Ethics* 2 (Spring), 11–32.
25. Plant, R.: 1977–78, 'Gifts, Exchanges and the Political Economy of Health Care', Parts I and II, *Journal of Medical Ethics* 3, 166–173; 4, 5–11.
26. Ramsey, P.: 1970, *The Patient as Person*, Yale University Press, New Haven.
27. Schwartz, B.: 1967, 'The Social Psychology of the Gift', *American Journal of Sociology* 73, 1–11.
28. Sells, R. A.: 1979, 'Let's Not Opt Out: Kidney Donation and Transplantation', *Journal of Medical Ethics* 5, 165–169.
29. Shelp, E. E.: 1981, 'Justice: A Moral Test for Health Care and Health Policy', in E. E. Shelp (ed.), *Justice and Health Care*, D. Reidel Publ. Co., Dordrecht, Holland; Boston, U.S.A., pp. 213–229.
30. Siegler, M.: 1979, 'A Right to Health Care: Ambiguity, Professional Responsibility and Patient Liberty', *The Journal of Medicine and Philosophy* 4, 148–157.
31. Silver, M.: 1980, *Affluence, Altruism and Atrophy*, New York University Press, New York.
32. Titmuss, R. M.: 1971, *The Gift Relationship*, Pantheon Books, New York.
33. Veatch, R. M.: 1976, 'What is a "Just" Health Care Delivery', in R. M. Veatch and R. Branson (eds.), *Ethics and Health Policy*, Ballinger Publ. Co., Cambridge, Mass., pp. 127–153.
34. Wilensky, H. L. and Lebeaux, C. N.: 1958, *Industrial Society and Modern Welfare*, Russell Sage, New York.

EARL E. SHELP

EPILOGUE

The imperatives of the principle of beneficence seem particularly relevant to the enterprise of health care in which relief from disease and enhancement of the human condition are central concerns. The relationship of the principle to health care issues has been partially explicated in this collection of essays. Insights into the relevance of the principle to medical-moral issues on an individual and social level have been provided. The historical perspectives, conceptual inquiries, and theoretical analyses presented here combine to provide, at least, a starting point, and perhaps a framework for additional research into this vital area of interest.

Darrel Amundsen and Gary Ferngren have shown that medicine has conceptualized and expressed in various ways the counsels of beneficence. This diversity of representation may reflect not only the particularities of the practice of medicine but also the theoretical uncertainties and inadequacies described by Allen Buchanan and William Frankena. On a more practical level, John Reeder reminds us that the nature and limits of the duties of beneficence remain a matter of debate.

The traditions of religious ethics reviewed by Ronald Green, William May, and Harmon Smith provide additional resources to be drawn upon in shaping the contemporary discussion of the relationship of the principle of beneficence to health care. The dynamic history of Jewish and Roman Catholic interpretation of the principle and its relevance to health care illustrates both the problems inherent to the principle and its promise for instruction in morally perplexing situations. The contribution of Protestant ethics may be to assist us in living with moral ambiguity or alternately cause us to probe the depths of moral wisdom to discover and agree upon those ends which are proper to human existence and human activity.

Given this state of theoretical uncertainty, four authors addressed the application of the principle of beneficence to a variety of specific concerns in health care. The meaning and scope of beneficence was suggested by Earl Shelp and Natalie Abrams respectively. Its application to matters of health policy was commented upon by James Childress and Ronald Green. The principle contributes to our understanding of the moral conduct of

Earl E. Shelp (ed.), Beneficence and Health Care, 255–256.

health care. But an appreciation for the problems of its interpretation and application cautions against an appeal to it alone.

Studies of moral theory need to redress the lack of attention given to this important principle of morality. Its apparent relevance to issues in health care mandates that its resources no longer remain latent and that its contribution to health care become more explicit. A more complete understanding of the principle and its implications for issues in health care are provided in these essays. With a sustained investigation of this essential moral relationship and its components, our appreciation for and understanding of beneficence and health care should be enriched.

Institute of Religion and
Baylor College of Medicine
Houston, Texas

NOTES ON CONTRIBUTORS

Natalie Abrams, Ph.D., is Assistant Professor of Philosophy and Co-Director, Philosophy and Medicine Program, New York University School of Medicine, New York, New York.

Darrel W. Amundsen, Ph.D., is Professor of Classics, Western Washington University, Bellingham, Washington.

Allen E. Buchanan, Ph.D., is Associate Professor of Philosophy, University of Minnesota, Minneapolis, Minnesota.

James F. Childress, Ph.D., is Professor of Religious Studies, University of Virginia, Charlottesville, Virginia.

Gary B. Ferngren, Ph.D., is Associate Professor of History, Oregon State University, Corvallis, Oregon.

William K. Frankena, Ph.D., is Emeritus Professor of Philosophy, University of Michigan, Ann Arbor, Michigan.

Ronald M. Green, Ph.D., is Associate Professor and Chairman, Department of Religion, Dartmouth College. He is also an adjunct member of the Department of Community and Family Medicine, Dartmouth College Medical School, Hanover, New Hampshire.

William E. May, Ph.D., is Associate Professor of Moral Theology, The Catholic University of America, Washington, D.C.

John P. Reeder, Jr., Ph.D., is Professor of Religious Studies, Brown University, Providence, Rhode Island.

Earl E. Shelp, Ph.D., is Associate Professor of Theology and Ethics, Institute of Religion, and Assistant Professor of Ethics, Baylor College of Medicine at the Texas Medical Center, Houston, Texas.

Harmon L. Smith, Ph.D., is Professor of Moral Theology, Divinity School, Duke University, Durham, North Carolina.

Earl E. Shelp (ed.), Beneficence and Health Care, 257.

INDEX

The Philosophy and Medicine Book Series

Managing Editors

H. Tristram Engelhardt, Jr. and Stuart F. Spicker

1. **Evaluation and Explanation in the Biomedical Sciences**
   1975, vi + 240 pp. ISBN 90-277-0553-4

2. **Philosophical Dimensions of the Neuro-Medical Sciences**
   1976, vi + 274 pp. ISBN 90-277-0672-7

3. **Philosophical Medical Ethics: Its Nature and Significance**
   1977, vi + 252 pp. ISBN 90-277-0772-3

4. **Mental Health: Philosophical Perspectives**
   1978, xxii + 302 pp. ISBN 90-277-0828-2

5. **Mental Illness: Law and Public Policy**
   1980, xvii + 254 pp. ISBN 90-277-1057-0

6. **Clinical Judgment: A Critical Appraisal**
   1979, xxvi + 278 pp. ISBN 90-277-0952-1

7. **Organism, Medicine, and Metaphysics**
   Essays in Honor of Hans Jonas on his 75th Birthday, May 10, 1978
   1978, xxvii + 330 pp. ISBN 90-277-0823-1

8. **Justice and Health Care**
   1981, xiv + 238 pp. ISBN 90-277-1207-7

9. **The Law-Medicine Relation: A Philosophical Exploration**
   1981, xxx + 292 pp. ISBN 90-277-1217-4

10. **New Knowledge in the Biomedical Sciences**
    1982, xviii + 244 pp. ISBN 90-277-1319-7